Just-in-Time Systems

P9-AEV-860

DISCARD

Uwe Holl is a member of the executive board of the Institut der Deutschen Wirtschaft in Cologne, responsible for information, documentation and international relations.

Dr. Malcolm Trevor is Chairman of the Euro-Japanese Management Studies Association at the Policy Studies Institute London, Britain's largest independent research institute for the study of economic, industrial and social policies and the institutions thereof. He previously edited *The Internationalization of Japanese Business* (Campus/Westview 1987)

Uwe Holl, Malcolm Trevor, Editors

Just-in-Time Systems and Euro-Japanese Industrial Collaboration

Campus Verlag · Frankfurt am Main
Westview Press · Boulder, Colorado

Copyright © 1988 in Frankfurt am Main by Campus Verlag

Published in 1988 in the United States by WESTVIEW PRESS
Frederick A. Praeger, Publisher
5500 Central Avenue
Boulder, Colorado 80301

ISBN 3-593-33970-6 Campus Verlag
ISBN 0-8133-0729-5 Westview Press

Library of Congress Cataloging-in-Publication Data
Just in time systems and Euro-Japanese industrial collaboration.
1. Production control. 2. Inventory control.
I. Trevor, Malcolm, 1932-   . II. Holl, Uwe.
TS 157.J87   1988   658.5   88-17100
ISBN 0-8133-0729-5

CIP-Titelaufnahme der Deutschen Bibliothek

*Just-in-time systems and Euro-Japanese industrial
collaboration* / Uwe Holl ; Malcolm Trevor, ed. – Frankfurt am
Main : Campus Verlag ; Boulder, Colorado : Westview Press,
1988
    ISBN 3-593-33970-6 (Campus) kart.
    ISBN 0-8133-0729-5 (Westview Press) kart.
NE: Holl, Uwe [Hrsg.]

All rights reserved. No part of this book may be reproduced or transmitted in any form
or by any means, electronic or mechanical, including photocopying, recording, or by any
information storage and retrieval system, without permission in writing from the
publishers.
Printed in West Germany

TABLE OF CONTENTS

FOREWORD

In taking up the two issues of just-in-time systems and Euro-Japanese industrial collaboration together, we are probably breaking new ground. Both are complex, forward-looking topics. Each has often been discussed separately, but I am not aware that both have previously been treated together under one heading.

At first glance they may seem to have nothing to do with one another, but do they not have something in common? The fact is that both have gained in importance in recent years, not only in scope but also in character. This increase in importance, which has occurred in different ways in different countries, is itself justification for the papers and discussion that follow. On the other hand, it is clear that the Anglo-German Foundation, which provided the initial impetus, had something more in mind when they proposed this composite question as a subject for dicussion.

Traditionally, for instance, competition and cooperation have been seen as alternatives. But the internationalisation of markets and production has increasingly made them complementary. Today enterprises that compete with one another are at the same time forced to cooperate and therefore they need to have a knowledge of the infrastructure of the firms in the different countries.

This process is increasingly being expedited by the introduction of just-in-time (JIT) systems, especially in Japan. The example of Toyota shows that the JIT concept is not just a means of rationalisation or of increasing productivity but something that extends beyond production and sales or marketing into all areas of decisionmaking.

In Japan, the Advisory Group on Economic Structural Adjustment for International Harmony submitted a report, widely known as the Maekawa Commission Report, to Prime Minister Nakasone in April 1986. Among other things, it stated that, "Industrial cooperation, including technology transfer and cooperation in third country markets, should be actively pursued. In particular, the establishment should be promoted of an institution from the private sector, whose goal should be the exchange of personnel within the framework of industrial cooperation."

A proposal of this type offers a means of increasing the number of managers with international experience. Such people are needed today to initiate and extend cooperation between enterprises in different countries. For the introduction and implementation of JIT systems, managers with ideas, and -- if possible -- both local and international experience are particulary necessary at all levels.

As far as the strengthening of Euro-Japanese industrial collaboration is concerned, one of the most important preconditions is that all the parties involved should learn to trust each other more and be more prepared to enter into long-term commitments. These would also appear to be the most important preconditions for the efficient implementation of JIT systems in the supplier-customer relationship, especially when this cuts across national boundaries - whether in terms of communication and the flow of information, the flow of materials and products, cost reduction, or the maximisation of marketing strategies.

Finally, both cooperation that extends beyond national boundaries and the implementation of JIT systems impinge on the relations between employers and unions. Companies become more vulnerable to strikes, and unions are in a position to pursue strikes with greater effect in proportion to the resources that they themselves commit. We are therefore grateful for the participation of British and German trade unionists in the following discussion.

I am personally convinced that a better understanding of the JIT systems that have been introduced in Japan, Britain, and Germany and of their implications will have a positive effect on broadening and deepening collaboration among the three countries -- perhaps not only in the industrial field but also in the field of trade. The composite nature of the questions posed by the title is therefore particularly farsighted, and I would like to thank the Anglo-German Foundation for the Study of Industrial Society for taking them up and enabling us to hold a conference where we were able to discuss them from the perspective of practical experience.

This is not the first time in its relatively short existence that the Foundation, which has already sponsored many projects of real practical value to all those involved, has demonstrated its ability to see beyond the formal limitations of a topic and to look to the future through its policy of actively promoting discussion. The Foundation had indeed already taken up the triangular relationship between Japan, Britain, and Germany once before, when it sponsored a study (led by Dr. Malcolm Trevor) of the Japanese management development system in the Federal Republic of Germany and Great Britain. This study found, among other things, that Japanese managers were more task-oriented than their British and German colleagues and that the latter were by comparison more individualistic, specialised, and inflexible. And this may explain something about the relationship between Japan and Europe. Indeed one reason why JIT systems have been introduced both more widely and more efficiently in Japan than in Britain or Germany may lie in the specific features of Japanese managerial practice just alluded to. Thus the management development study of 1984-85 has interesting implications.

We are extremely grateful to the Anglo-German Foundation for sponsoring the conference on JIT systems and Euro- Japanese industrial collaboration. It was held at the Institut der deutschen Wirtschaft, Cologne, in September 1986. This conference provided the occasion for a practical exchange of experiences and views in regard to a concept which, if correctly and consistently applied, is capable of mobilising powerful forces of rationalisation and productivity for the benefit of us all.

I would like to thank the Japan External Trade Organisation (JETRO) and particularly its Deputy Director, Mr. Shoji Isaki (the text of his presentation at the conference is given below) and the Japanese Chamber of Commerce and Industry in Düsseldorf and its general manager, Mr. Akira Arikawa. We are most grateful to both organisations for their valuable help in bringing the conference to fruition.

Hans-Josef Breidbach
Deputy Director
Institut der deutschen Wirtschaft
Cologne

INTRODUCTION

JIT systems have many different aspects: technical aspects, social and organisational aspects, financial aspects, and last but not least, quality aspects. Obviously there is no point in supplying things just in time if the quality is not going to be right, and one manager in Japan was quoted as saying that one of the points of JIT was actually to induce a permanent sense of crisis. It is a system in which there is no slack, and this is a way of keeping people on their toes, of motivating them, and of course of getting maximum performance. So it is very, very demanding.

In Britain, we now have two Japanese electronics components manufacturers, Alps Electric and Tabuchi. In the United States also, there are now quite a few Japanese motor components makers such as Stanley Electric Company and others who are producing locally and supplying not merely the Japanese motor manufacturers but also Ford, General Motors, and Chrysler, or the "Big Three" of the U.S. motor industry. Some people are wondering whether this trend will be repeated in Europe and perhaps particularly in Britain with the arrival of Nissan.

Obviously, JIT is a very topical subject. The system introduced at Lucas has recently been described in Management Today. Philips has introduced the concept of co-makership, or close relations between customer and suppliers. Nissan and Honda in the UK have assented to British government urging that they attain a local content of at least 80 percent (which is a very high figure), by the early 1990s. Nissan is taking a long-term view on parts suppliers, and 27 British component manufacturers together with three from other EC countries have now passed Nissan's stringent quality and delivery standards. Some of these things have even been shown on television in the UK, and another Japanese company that is starting manufacturing in Britain, Komatsu, is also giving contracts to local suppliers. So JIT is certainly a topical issue, and we are fortunate to have presentations from the two sides of the customer-supplier relationship.

It is therefore a special pleasure to the Anglo German Foundation for the Study of Industrial Society to have sponsored a conference that so many practitioners found the time to attend. We feel the conference proceedings presented in this book may be of some value to businessmen on both sides of the Atlantic whose responsibility it is to stand up to international competition. But today we often find a tendency to be parochial and self-absorbed in looking at each other. That, I think, is a bit limiting, and so we were particularly grateful to the Japanese participants for holding up a mirror to us. This helps us to look at each other in the Japanese reflection and to see that some things that seemed impossible in European eyes suddenly seem to have been made possible in Japan.

Before we get too worried about this, I still remember the U.S. challenge when Europe did not seem able to get itself organised, and the Americans were walking all over us. But we learned to cope with the U.S. challenge in the various countries -- at least in Germany. So I am sure our Japanese friends will not take offence if we say that we will one day catch up with their industrial advances, too. Only I suspect we will not catch up then as Britain and as Germany, but as a much more integrated Europe.

Hans B. Wiener
Project Director
Anglo-German Foundation for the
Study of Industrial Society
London

EURO-JAPANESE INDUSTRIAL COLLABORATION:
EXPERIENCES AND PROSPECTS

Shoji Isaki
Deputy Director General
JETRO
Düsseldorf

JETRO was founded in 1958 under a special law by the Japanese government, which has a 100 percent share of the capital. Its activities are mainly financed by the Japanese government. The Japan Export Trade Research Organisation has used the name "JETRO," since 1951, before the present statute was enacted. This organisation was the foundation on which the present JETRO was established. In fact, the name JETRO itself derives from BETRO, meaning British Export Trade Research Organisation, which was very successful at that time in promoting British exports. Although JETRO had no direct institutional contacts with BETRO, Japan's high regard for BETRO was the motivation for establishing the same kind of institution. In this sense, we owe much to Great Britain as our predecessor, and if we consider that BETRO had a great influence on the establishment and activities of JETRO, it might be said that the establishment of JETRO was a kind of cooperation between Great Britain and Japan.

BETRO does not exist any more, but JETRO, under its official title of Japan External Trade Organisation has developed its activities in accordance with the changing needs of the times. Until about 15 years ago, JETRO's most important activity was the promotion of Japanese exports. But as Japanese foreign trade began to record a bigger and bigger surplus, it became JETRO's most important task to promote imports from foreign countries and to promote industrial collaboration in the form of capital and technology exchange so that JETRO can contribute more effectively to the revitalisation and prosperity of the world economy. JETRO's main activities in the field of promoting investment abroad include:

a) Putting on file Japanese enterprises that have interests and investments abroad,

b) Publishing the list of such enterprises,

c) Assisting foreign organisers of seminars or delegations to Japan in connection with the promotion of Japanese investment in their countries,

d) Consulting on overseas investment for Japanese companies.

Japanese Direct Overseas Investment

According to JETRO's 1986 estimates, the stock of foreign direct investment of the six major industrialised countries (the USA, the United Kingdom, Japan, the Federal Republic of Germany, Canada, The Netherlands) reached US $ 535 billion at the end of 1985. On the basis of this amount for the six countries, the world's total outstanding direct investment is estimated to have increased by 16.3 percent in 1985 and to have reached US $ 644.6 billion. Of the total for the six major countries, the US holds 42.5 percent, the UK 15.5 percent, followed by Japan with 6.9 percent, and West Germany with 6.7 percent. Japan has therefore now moved into third place in world foreign investment.

Japanese direct overseas investment started to increase for the first time in the second half of the 1960s and reached US $ 3.5 billion in fiscal 1973. After the 1973 oil shock, Japanese overseas investment entered a period of stagnation for some years but started to increase again in 1978 and reached US $ 12.2 billion in fiscal 1985. The appreciation of the yen indicates a trend for Japanese direct foreign investment to continue increasing in the future.

Looking at the past development of Japanese investment abroad, there are some noticeable changes in the pattern of investment in the 1970s and 1980s. While the total stock of Japanese investment increased 5.6 times in the period 1974-1984, the stock of investment in manufacturing industry increased 5.3 times, and in non-manufacturing industries 5.8 times. As a result, investment in non-manufacturing industries accounted for a share of 70.8 percent at the end of fiscal 1985. In manufacturing industry, there has been a shift in importance from textiles, woodworking, and pulp to heavy industries such as iron and steel, metalworking, chemical, electric, and transportation machinery industries, and this tendency became even more obvious in the 1980s. There have also been changes in the regional distribution of investment, with an increasing share of investment in industrialised countries especially in

14

manufacturing. The portion of Japanese investment in foreign manufacturing, that went to the USA increased from 16.7 percent in 1974 to 29.4 percent in 1984, and the amount that Europe received increased from 5.2 percent to 8.0 percent.

Summing up, it may be said that the share of Japanese direct overseas investment in developing countries that goes to natural resources and labour-intensive industry is decreasing, whereas the share of Japanese investment of industrialised countries that goes to non-manufacturing sectors and technology-intensive manufacturing is increasing and is expected to increase further in the future. The increase in Japanese investment in the manufacturing sector of industrialised countries in the 1980s was mainly caused by the increase in such investment in the US and Europe. The reasons can be found in the facts that:

a) the difference in labour costs between Japan, the US, and Europe is becoming smaller;

b) the political situation in both regions is stable, and access to related industries, capital, and information is relatively easy;

c) the respective governments are willing to stimulate their industries through foreign investment.

Of course, Japanese enterprises are motivated in the sense that in recent years, they have had greater resources in the areas of technology, capital, sales networks, and management, and they are developing international activities not only in terms of product exports but also in terms of local production.

The above is a review of some of the factors leading to an increase in Japanese companies' investment in manufacturing in developed countries. But, as to the nature of such investment, to some extent it cannot be denied that investment in manufacturing has been increasing with the aim of countering or completely preventing the increasing danger of protectionism.

In accordance with the growing interest of European business circles in Japanese direct investment in Europe and in Japanese companies, management techniques, JETRO conducted two studies of Japanese production companies in European countries. One of these was done in 1985 with the purpose of investigating the state of labour relations and the procurement of parts and materials for local production. So what follows is an outline of the main results of this study in the area of parts and materials procurement.

15

The study was conducted by sending questionnaires to the managers of 188 Japanese companies known to be producing in Europe at that time. Out of 188 companies, 119 answers were received.

Before coming to the subject of the procurement of parts and materials, it is necessary to explain the distribution of these Japanese companies so that the development of the production activities of Japanese companies in Europe can be better understood.

1. Timing of the Establishment of Japanese Production in Europe

Table 1 below shows that the level of investment has been steadily stepped up as expectations for industrial cooperation between Japan and Europe have risen.

Table 1: Date of establishment of companies

	No. of firms established in Europe
pre- 1970	18
1971–1975	54
1976–1980	51
1981–1984	65
Total	188

2. Distribution by Country and Sector

The affiliates of Japanese companies are concentrated in major EC nations such as West Germany (34), the UK (32), and France (30). Until some years ago, France was a country where foreign investment was relatively difficult because of government restrictions. But now, there are no restrictions any more. That is one point, but another important point is the fact that it is difficult in France, even after the liberalisation of foreign investment, to invest in the service sector, including trade and exports, and to expand the sales networks for Japanese or foreign products. The French government is willing to welcome manufacturing investment, but it is difficult to establish a good sales network because investment in that field is not so easy, and there is a kind of confrontation. Japanese manufacturers are in some sense forced to establish production in France and to contribute to employment there. One Japanese manager said the background was that we were always trying to promote Japanese investment in European countries so there would be more industrial collaboration between Japanese and European industries. But he also said the background is, of course, the big surplus in Japanese exports to Europe; so we believe Japanese production should also be started in Europe. The manager said it is difficult to start production in France as long as there is a big surplus in Japanese exports to France. That is, if the surplus in Japanese exports to France were smaller or if exports and imports were balanced, investment in commercial areas would be easier. Then Japanese companies might be able to sell more products there and start production, too. So, there is a contradiction.

By industry, there were 52 companies in the high tech industries, electronics and electrical equipment, with a share of almost 30 percent of the total. Next to those in the electronics and electrical equipment sector were companies classified as "others," including slide fasteners, lenses, small sealed batteries, and packaging materials. The products manufactured in the electronic and electrical equipment sector included integrated circuits, video recorders, colour television sets, hi-fi systems, and semi-conductors. These products are fairly high in added value and are, indeed, items that generate conspicuous trade friction, or are liable to cause friction in the future.

3. Scale of Companies

The total capital of the 119 responding companies was US $ 493.22 million, with an average of US $ 4.14 million per company. The Japanese affiliates in Europe may be said to be big businesses in so far as capital is concerned, but a look at the number of employees shows that over 80 percent of the local operations were "small and medium size businesses" with less than 300 employees (see Figure 1 overleaf).

17

Figure 1

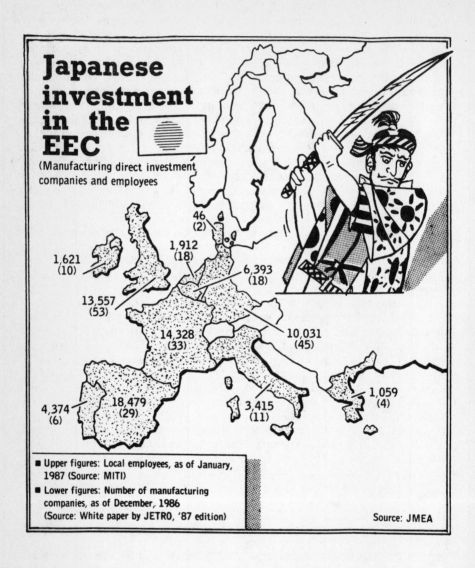

Japanese investment in the EEC

(Manufacturing direct investment companies and employees

46
(2)

1,912
(18)

6,393
(18)

1,621
(10)

13,557
(53)

14,328
(33)

10,031
(45)

1,059
(4)

4,374
(6)

18,479
(29)

3,415
(11)

- Upper figures: Local employees, as of January, 1987 (Source: MITI)
- Lower figures: Number of manufacturing companies, as of December, 1986 (Source: White paper by JETRO, '87 edition)

Source: JMEA

Source: The Guardian, 15 April 1987

4. State of Operations and Prospects for Future Operations of Companies

Of the 119 companies responding to the survey, there were 57 subsidiaries owned 100 percent by Japanese companies (48.7 percent) and 56 joint ventures with local capital (47.8 percent). The remainder consisted of further subsidiaries of Japanese local subsidiaries.

As to future prospects, 97 out of the 119 responding companies, equivalent to 82 percent, made optimistic projections. In particular, a large number of companies in basic resource type industries made optimistic projections, with nine companies responding, "will increasingly expand" and nineteen responding, "will grow fairly well;" close to 88 percent of the total. So far as the future prospects for operations were concerned, there was no great correlation with the country or sector invested in. Moreover, almost all the Japanese affiliates in Europe were confident about their future operations.

Local Procurement of Parts and Supplies: Problems and Prospects

1. Rate of Local Procurement of Parts and Supplies

In the study, questions were asked about the rates of local procurement of parts and supplies for the main products of the Japanese affiliates (the percentage by value of locally procured parts and supplies compared to the value of factory shipments). But there were few valid responses, so it was difficult to identify precise trends in the Japanese affiliates as a whole. However, the results showed a trend among Japanese affiliates to raise local procurement rates. For example, the average rate of local parts procurement was 74.8 percent at the time of establishment (in the case of 23 responding companies), but is now 80.2 percent (in the case of 29 companies). The rate of local procurement for special-order parts rose from 56.6 percent (20 companies) to 71.3 percent (20 companies). For supplies it rose from 79.8 percent (27 companies) to 89.5 percent (39 companies). All of this indicates considerable growth. The electronic and electrical equipment manufacturers raised their local procurement rates to a larger extent than the average partly because there were strong

19

pressures to increase local procurement rates under EC coun-
try-of-origin regulations.

2. Rate of Local Procurement by Size of Company

A look at the situation by size of company showed that 17 companies, or 40 percent of the 42 responding companies with 200
or more employees, were raising their rates of local procurement. Further, 17 companies, or 60 percent of the 28 responding companies with sales of over US $ 20 million, were raising
their rates -- a considerably higher proportion than for the
group of companies of smaller size. The larger the size of a
Japanese affiliate, the higher the rate of local procurement of
parts and supplies, it seems. And the rate of increase of local
procurement is higher in the case of general parts and special
order parts and lower in the case of material supplies. In the
latter case, it should be considered that companies had
already reached the maximum levels of local procurement of
supplies at the time of their establishment.
 Seen in this way, the rate of local procurement of parts and
supplies by Japanese affiliates has been increasing, primarily
among large companies. Looking at the reasons for the increase, the reason given most often for general parts, special
order parts, and supplies in common was that "an appropriate
source had been found for local procurement" (64 out of a
total of 111 responding companies for general parts, special
order parts, and supplies). Next to that, the most common
explanation was that "an appropriate local company able to
handle processing or outside orders had been found" (22 out
of the same 111 companies). Other reasons given were:

a) active purchasing of local parts and supplies due to de-
   precation of the pound sterling;

b) establishment of a local subsidiary for procurement of local
   parts and supplies;

c) rise in the quality of supplies from local manufacturers
   and thus switchover to local procurement;

d) costs and complexity of import procedures, making local
   procurement unavoidable despite the lower quality of locally procured supplies, etc.

So far as the results of this survey show, it is very important for increasing local procurement that Japanese affiliates should be able to find local manufacturers who can maintain high standards in producing, delivering, and processing parts and supplies according to the specifications of the assemblers. On the other hand, companies indicating that they were reducing their local procurement of parts and supplies were extremely few in number.

3. Size of Local Supplying Companies

So far as the survey results indicate, the companies' Japanese affiliates deal with for local procurement are overwhelmingly medium and small businesses. The reasons most often given for this were that, regardless of whether they were supplying general parts, special order parts, or other supplies, "The (medium and small-sized) companies were in a position to deliver under appropriate conditions in respect of quality, price and delivery" and that, "There were no appropriate large enterprises making the parts or supplies required."

4. Parts and Supplies Procurement Methods

Looking at the forms of procurement of parts and supplies, 41 of the 104 responding companies indicated that they engaged "principally in spot purchases," and 63 said they "principally used supply contracts for fixed periods." There were more companies operating on a fixed term basis.

It is rather difficult to explain why the percentages for spot purchasing (39 percent) and for purchasing for a fixed time (61 percent) have developed in this way. The life cycle of the products of the Japanese manufacturers here seems to be rather short, especially in the field of electronics, and that may be the reason for the high percentage of purchasing on the spot. For instance, the quality of German dies for the production of plastic parts and for the housing of some equipment is very high. These dies can be used for many years and for thousands of plastic parts, but a Japanese manufacturer said that they do not need such a long life for the dies because there may be a model change after only six months. They do not produce in such great numbers.

5. Return of Goods to Supplier

Japanese affiliates also have problems with local components manufacturers and suppliers in respect of delivery time, quality, and delivery terms. There is, without question, a large difference between Japanese manufacturers and European manufacturers with regard to strictness of delivery time, quality, and delivery terms. In the present survey, 68 out of 100 responding companies indicated that they had had to return goods to the supplier due to the problems mentioned.

There were a lot of cases of returning goods in all European countries and high rates of local parts and supply procurement. For example, 14 of the 17 responding companies in the UK and 7 of the 8 responding companies in Belgium indicated they had returned goods at least once in the past.

A look by industry shows that 24 of the 30 responding companies, or 80 percent, in the field of electronic and electrical equipment had returned goods -- far more than the 60 percent for companies in the field of general machinery, basic resource industries, and other manufacturing industries. Further, the larger the company, the more frequent the experience with returns. Specifically, 32 (or more than 80 percent) of the 38 companies with 200 or more employees and 24 (or just under 90 percent) of the 27 companies with sales of US $ 20 million or more indicated that they had returned goods procured locally.

Looking at the measures taken by Japanese affiliates to deal with such problems, 67 of the 114 responding companies (multiple responses included) indicated that they had actually returned the goods, 26 said they had claimed damages, and 10 had cancelled their contracts. Eight mentioned other measures such as accepting replacements for defective goods in some cases. Japanese affiliates have been taking quite thorough-going measures to prevent such problems. First, they check the competence and the attitude of local parts manufacturers' and other suppliers' management in advance. Then they provide technical guidance in production technology and quality control to the manufacturers and strengthen their own pre-delivery inspections. In particular, to press their requests and desires more effectively, some Japanese affiliates engage in detailed discussions with local manufacturers on technical matters and otherwise work to improve communication.

These facts indicate just how interested Japanese affiliates are in improving product quality. Many of the companies are ready to provide technical guidance to local manufacturers towards this end, and by these means alone it should be possible for Japanese affiliates to increase their procurement of local parts and supplies. In this sense, even greater efforts must be made to foster local parts manufacturers and suppliers who are able to meet the quality standards of Japanese affiliates.

Although Japanese affiliates face various problems in connection with local procurement of parts and supplies, about 40 percent of all the companies predicted their local procurement rates would rise in the future. Japanese affiliates can therefore be said to have a fairly optimistic outlook on improving their procurement of local parts and supplies, and from the viewpoint of industrial cooperation this will be mutually beneficial in many ways.

6. Problems with Sub-contracting

One way for Japanese affiliates to ensure a stable supply of parts at a certain quality level is to provide technical guidance or to lend production equipment to local manufacturers and thus to develop a mutually beneficial relationship between the parts supplier and the assembler. This amounts to the type of relationship between sub-contractor and assembler that is widespread in Japan but is not generally known in Western Europe. There is a double layer of industry in Japan. The large companies tend to have small and medium-sized companies balanced against them, and the difference in strength tips to the advantage of the larger companies.

On the other hand, when small and medium-sized companies in Japan depend upon large companies, they share the risk (because of the employment relationship in the country), the personal relationships, or the sub-contracting arrangement, so the large company places orders with them because of the set-up. Even when business conditions are not favourable, such relationships can be set up for small and medium- sized companies. They gain stability and can expect to enjoy stable continuity.

In this respect, German small or medium-sized companies are different. They are more independent and do not, for example, depend on a particular company for 30 percent of their business. So this difference in competition means there may be independence on the one hand, and on the other hand, you can have a good stable relationship with a supplier. So there is an advantage for the supplier in that.

Therefore, all cases where a Japanese company gives technical guidance, financial aid, management advice, or production machinery, are provisionally defined here as "sub-con-

23

tracting," and from this point of view it was interesting to see how Japanese affiliates were trying to establish sub-contracting relationships in the context of differing European business practices.

With regard to the main products of 79 responding companies, the average share of parts produced by local sub-contractors per value of factory shipments was not more than 9.3 percent, and only 24 out of 103 responding companies (or 23 percent) had sub-contractors. With the exception of Belgium, the share of companies that had local sub-contractors was around 20 percent in all cases. Viewed by industry, a relatively high proportion of companies in the electronic and electrical equipment field had local sub-contractors, reflecting the tendency for sub-contracting to be concentrated among large enterprises, particularly in the electronic and electrical field.

In actuality, 16 of the 35 companies with 200 or more employees (or almost 46 percent) and 16 of the 30 companies with sales of US $ 20 million or more (or some 50 percent) replied that they would need local sub-contracting manufacturers in the future. Seen by industry, 19 of the companies in the electronic and electrical equipment field (or more than 60 percent) responded that they would need subcontractors -- a far higher proportion than in other fields.

Recently, a study was made of the production of videotape recorders (VTRs) by Japanese companies in Europe that are producing locally at about 10 factories. All these manufacturers started production only after 1982, and the volume of production has been rapidly expanding.

There are also good examples of cooperation between Japanese manufacturers and European producers in the case of J2T and others, and Japanese VTR manufacturers have been steadily increasing local content according to the guidelines set out by the EC Commission. The present minimum local content set for Japanese manufacturers by the EC Commission at the end of 1985 is about 45 percent, and we will have to see if this minimum ratio of local procurement will be higher in the next few years.

Local parts procured in Europe differ from company to company, but Japanese companies have so far had the same experience in the sense that they have had great difficulties finding appropriate sub-contractors who are able to undertake sub-assembly process work for printed circuit boards and decks with mechanical functions.

This means, in turn, that it is getting more and more difficult for them to increase local content as rapidly as before.

But, on the other hand, they are willing and ready to increase local content as far as possible. There is also at least one Japanese manufacturer in West Germany and in Great Britain that is buying mechanical decks with mechanical parts – which are important parts – from a European country. Of course there are also sub-contracting companies producing printed circuit boards for the production of VTRs. In this respect, many Japanese VTR manufacturers in Europe point out that there are still possibilities for increasing local procurement in the fields of plastic parts, pressed and cut metal parts, and moulded and forged metal parts.

With this situation in mind, the author of the report suggested that Japanese manufacturers in Europe and those who are planning to invest in Europe in the future should take steps to strengthen their connections with European suppliers so they will be able to form a wide network of cooperation involving different aspects of the whole of local society.

In this respect, I agree with the author of the report and hope that collaboration among us will be further strengthened. As a JETRO study pointed out recently: When Japanese companies come to Europe, each company tries to procure parts from Europe, but they are not merely buying components. On the technical side, we try to enter into collaboration with local companies. As for future trends, according to our study, such collaboration will be extended by many companies. That is their wish, of course, although between Japan and Europe there are differences in experience and practice in these respects. Compared with Japan, the introduction of such a system is not so easy, but the Japanese companies do not want to force the system on local companies in Europe. Wherever possible, they want to carry this out step by step and to make a contribution to the local scene.

TOSHIBA'S APPROACH TO PURCHASING

Peter Bayliss
Materials Director
Audio Visual Division
Toshiba Consumer Products (UK) Ltd.

Toshiba Consumer Products (TCP) was established in Plymouth in May 1981, to manufacture colour television receivers for sale to the UK market. In that first year of manufacture, we made 76,000 televisions in nine different models. In our first six years in business, we have grown to five times that modest beginning, introduced an export range, and now have a current product range of more than 50 different models.

In 1984, in addition to our performance in the colour television market, we established a new factory in Plymouth for the manufacture of microwave ovens. Although it is very young, this new factory will continue in the traditions and ideologies of the company we established more than six years ago.

In the UK, our production is all sold through our sister sales company Toshiba (UK) Ltd. located at Frimley in Surrey. Our export sets are marketed via Toshiba sales subsidiaries in West Germany, France, and Switzerland and through appointed distributors in other European countries.

The marketing of consumer durable products of the type we manufacture is one of the keenest in the UK and the rest of Europe, and there is continuing pressure to reduce prices and provide a high degree of service to our customers. In particular it is of the utmost importance that we have high quality products available at the right time and price.

In order to achieve this, we have to operate with the minimum lead time consistent with the procurement of materials and our production cycle time. Furthermore, the demand for our product is cyclic, with a high proportion of our production required during the last three months of the year. This requires us to undertake a degree of smoothing of our production and a policy of stockholding during the summer months.

One of the keys to success in business is attention to detailed planning which is subsequently subject to little or no change. In our relationships with our sister sales companies and our parent company in Japan, we have developed a routine monthly planning cycle of which we can justifiably be proud. The degree of change requested by our sales companies is virtually nil, and this has contributed to a great extent to our ability to work with a lead time as short as four months from planning to finished set production. However, they do occasionally ask us to change the schedule within that four months' lead time. If we are able, we will certainly do it, in terms of, for instance, colour variation on a cabinet or something like that.

But one should not get carried away too much by four months because that is not a four months' lead time for the supplier. That is four months from starting our plan to our actual production. We have administrative and processing time before that. We do have an in-house manufacturing time for small electronic components that are used very early on in the process. That probably means that the actual lead time the supplier will get will be down to about eight weeks, which is about 50 percent of that actual lead time. But we are probably slightly different from most other Japanese companies about lead times. I believe we are much more rigid in our plans than most other Japanese manufacturers in the UK.

In terms of our European customers, certainly Toshiba again has a good understanding. Distribution is much more difficult in those countries where we do not have Toshiba companies. There we get a good indication and to some degree make our own internal judgement of what to manufacture. It is not wholly satisfactory, but we can enlist the cooperation of all our sales companies to help us out. However, that does not mean we do not achieve a much greater degree of efficiency by maintaining very firm schedules.

With our UK company, we attempted to look right into the future, something like three years ahead. In more detailed terms, we attempt to look a year ahead. But in the short term, we have an understanding with our UK marketing company, and they totally understand our requirements and our lead time factors.

In Japan, lead time is two and a half months, coming down to something less than two months. They are now reviewing it. Our parent company would in fact like us to have the same sort of lead time. I am sure the market is demanding that you have a lead time of from today to tomorrow, but obviously we have to live in the real world.

There are tremendous market pressures on all suppliers of consumer products to shorten the inspection time of any new product. This means that the amount of time we can give to design information becomes shorter and shorter. We are obviously in a highly competitive industry, and it is difficult to have much confidence that we can increase that cycle time and give even more lead time in terms of design information. The market is moving very fast, and we just have to keep up with it and shorten our lead time as much as we possibly can to produce the product in time.

Figure 1 (see page 40) illustrates the monthly planning cycle. On the 15th of month N, the firm plan for production during the period 11th of month N + 4 to the 10th of month N + 5 is agreed. This plan, which is on a model-by-model basis, is sent by facsimile to Japan where it is entered into their computer and exploded into the constituent parts. The results of this computer operation in Japan now follow routes that align with our three major sources of parts:

1. Toshiba Corporation in Japan,
2. other Far East suppliers with whom we deal directly,
3. local suppliers (British and Continental).

For Toshiba parts, order details are sent directly to the divisions where the parts are made, for subsequent shipment to our overseas department for kitting.

For other direct Far East parts, an advance notification of order is sent. This is subsequently followed up by an official order from TCP, and we have an agreement with our suppliers that our official order will not deviate from the advance notification.

For our locally procured parts, this detail is sent from Japan to TCP in the form of a floppy disk. There it is entered into our own computer, and after further processing and checking, purchase orders are produced and sent to the suppliers.

The plan is input on a monthly basis and processed to calculate the detailed material requirements and print orders that are then sent to our suppliers. In fact, to gain time we transmit Far East order details via satellite link to our parent company in Tokyo for onward transmission to our direct suppliers. These are then followed up by our official orders, which of necessity have to go by post.

Ordering Policy

TCP started in a relatively small way with an average of six different batches per month, and we adopted a system of placing component orders monthly in accordance with our production batches as shown in Figure 2 (see page 41). This did mean that for a common component we placed six orders in each month. This might appear expensive in terms of administration, but it also meant we could receive our materials on a batch-by-batch basis and immediately on receipt of the goods, store them in a unique location by batch. This apparently simplistic mode of operation has served us well for four years, but as can be seen from Figure 3 (page 42), our very rapid growth has now produced a situation where we have had to review our procedures. This has come about because we are now averaging over 40 batches per month, which is putting a considerable strain on our already hard pressed resources. We have therefore recently changed from a batch-by-batch oriented procurement system to a part-by-part procurement system early this year. To overcome any problems we may have internally in TCP, such as loss or damage, we also order and maintain a small stock of contingency materials for each stock number. I must emphasise that this is to take account of any internal problems within TCP and not to overcome supplier shortcomings. We established a set of fundamental principles at the outset of TCP and have endeavoured to maintain these "tablets of stone" since the early days of the company.

Examples of such philosophies are:

1. We will not change our orders.
2. We will pay on time as long as parts are delivered on time.
3. We will pay correctly if parts are of good quality.

As we have grown, these fundamental planks of business philosophy have been maintained as far as possible; although as we have grown both in volume and complexity, it has clearly become more difficult.

What We Expect from Our Suppliers

In line with our own philosophies, we expect a number of reciprocal attitudes from our suppliers. We expect not to have to chase suppliers for delivery. Once an order is placed and accepted, we expect the goods to arrive on the specified day. We have a fundamental policy of no goods inwards inspection. We expect to receive good quality components at all times. We do not count any incoming materials. We expect exact quantities to be supplied at all times.

In terms of many traditional attitudes to supply, some of the above may appear to be harsh and to some degree very difficult or almost impossible to achieve. However, my own belief is that attention to this level of detail is one of the reasons for the success of Japanese companies today. Again, if the attention is given to the customers' requirements, there is a knock-on effect in the remainder of the business in that a similar level of detailed scrutiny is applied and will inevitably result in an overall improvement in business efficiency and operation.

On the question of acccurate quantities, in the early days of the company before our suppliers had become fully accustomed to our requirements, we did find that we had a number of shortages at the end of most batches. This meant we had to withdraw components from our contingency stocks. So we instituted a total count of one batch, which required us to count approximately a million and a half components.

The results, as in Figure 4 (page 43), were analysed and tabulated and showed some very interesting facts:

1. Far East suppliers were far more accurate than local suppliers.
2. If there was any tendency to supply incorrect quantities, then our Far East suppliers erred on the high side, whereas local suppliers' errors were equally spread about the required quantity.

Also, from this information, we prepared a league table of performance. Without giving away any information about other vendors, we showed our suppliers where they stood in this league table and discussed any ways in which they could improve.

One company in particular, which was near the bottom of the table, did respond in a most positive manner and showed a considerable improvement in a short space of time. This, they said, had largely been achieved by looking at their own internal procedures and disciplines and by making a number of radical changes: By adopting some of our own philosophies, they had improved their performance not only to us but also to their entire customer base. In fact, it enabled them to take on more business as all of their own customers saw a rapid im-

31

provement in performance and hence gave them more orders.

One major aspect where our suppliers have slipped since we started is adherence to delivery dates. We had a policy of not chasing suppliers, but we are finding that more and more of our suppliers are failing in meeting our required dates, and we are having to spend time in chasing them. Of course, this is totally non-productive for us. We have also had instances, particularly with suppliers of proprietary electrical components, when because of a problem at the suppliers' premises, we have been put on "allocation." Due to our "exact" ordering policy, this means that our own production lines will be stopped, and therefore this situation is untenable for us. It did take some time to get this message across to some of our suppliers, but it cannot be emphasised strongly enough that allocation cannot be tolerated.

Turning now to quality: Our policy of no goods inwards inspection requires 100 percent good-quality component deliveries. But we do not have our heads completely in the clouds, and we certainly understand that in the real world 100 percent perfection will not always be achieved: However it must always be aimed for.

Like the majority of Japanese suppliers, we do not operate an acceptable quality level (AQL) system which has the effect of setting a target but offers no incentive for improvement. We adopt a simple approach using a parts per million (PPM) philosophy in which PPM targets can be regularly reviewed. Once a target has been reached, a new target can be set to improve performance still further. As with delivery performance, in carrying out our ideas, suppliers have been able to improve their performance to their other customers. The majority have also commented that paying attention to delivered quality has improved their overall company efficiency. It certainly reduces the amount of totally non-productive time in providing replacements or spending time in rework operations that are inevitable with an "exact" ordering policy.

Sourcing Policy

We have two major sources of components: Far East and local. The Far East is further subdivided into two categories, Toshiba and direct suppliers. Direct suppliers are located in a number of countries in the Far East, including Japan, Singapore, Malaysia, and Taiwan.

In terms of sourcing, Figure 5 (page 44) shows the percentage purchased from the three main sources and how this has changed during the first four years of our operation. We do in fact make great efforts to increase our local purchase, and in financial terms, we have increased our percentage of local purchases from approximately 40 percent to 50 percent. In real terms, the increase is considerably greater than this, as the pound sterling has weakened considerably during these four years, and this has inflated our Far East purchase prices. So if we relate back to a common exchange rate, then the increase in local purchase is from 40 percent to 61 percent (see Figure 6 on page 45). We do in fact have a localisation programme, which is continually in operation, to investigate possible new sources of components consistent with price, quality, and delivery performance. There are obvious benefits to us in so doing:

1. protection against fluctuating exchange rates,
2. avoidance of 30 days' shipping time and approximately two weeks customs clearance time,
3. faster response time if problems occur,
4. avoidance of freight and duty.

There are also drawbacks that make it difficult to achieve complete localisation or indeed to proceed as fast as one would like:

1. Our designers are all in Japan, and there is an obvious tendency for them to home in on their own components.
2. It takes time and resources to approve any new component, and as this has to be undertaken by the design authority, this activity is necessarily done in Japan.
3. In many instances, Japanese and indeed other Far East countries' components are considerably cheaper than local equivalents. In fact, in some instances, there is still a gap to be bridged even when the cost of freight and duty is added to the Far East cost.

The Question of Cost

Of all the considerations taken into account when purchasing, the one that is given more time than any other is normally the cost. While quality and delivery performance are equally important, the supplier is really expected to achive these two attributes.

33

Component cost has to be negotiated in line with a number of factors that are not mutually independent of one another, such as: the finished product market, competitiveness with other suppliers, volumes, transport, past performance.

All have a marked effect on what one is prepared to pay. It is also very often the case that a Japanese company will give a target price that the supplier must meet. The word inflation has virtually been struck out of our vocabulary: We expect, like other Japanese manufacturers, to receive a cost reduction at least once and if possible twice a year. Perhaps the very nature of our business demands that this is the case. Ten years ago, the average price of a 20-inch television was approximately 320 pounds sterling, while the average wage was about 30 pounds per week, i.e. a television cost about as much as three months' salary. Today, the comparative price is about 230 pounds or less against an average weekly wage of 140 pounds, or less than two weeks' salary.

Last year, our cost reduction targets were of the order of one million pounds. These were achieved 100 percent within just six months. It was also interesting to note that 1985 was the year in which we saw cost reductions provided by local suppliers exceeding those from Far East vendors, although Far Eastern companies are more aggressive as regards price than any other companies in the world.

We ourselves actually changed our organisation in order to achieve our objectives. We split the purchasing function into our three major origins of components (TSB, direct Far East, and local) and each of the three sections was responsible for all aspects of the job, namely: prices, buying, scheduling, stock control, storage, etc.

The problem was that each section was so busy ensuring the parts were at TCP in time for production that they had no real time for price negotiation and cost-down. So we changed the organisation, making the procurement section wholly responsible for the day-to-day activities of scheduling and ensuring timely delivery of parts. This allows the three purchasing managers to devote much more of their time to price negotiation and the introduction of new parts.

The recent weakness of the pound sterling has now highlighted a need to look towards 1987. In March 1986, the exchange rate was 260 yen to the pound. In September 1986, it was approximately 230 yen to the pound, resulting in our Far East components costing some 12 percent extra. Taking into consideration other factors such as salary increases and overhead costs, we somehow have to save more than 6 million pounds to ensure our survival in 1987.

We have therefore instituted a major project that we call "score" to look at all aspects of our business such as material costs, value analysis activities, overheads, productivity, and freight and storage costs. This project will involve not only TCP but our parent company in Japan and also our suppliers, as it is fundamentally important to all of us to succeed. We shall not wait until the problem is upon us, but by adopting this approach some nine months before the start of our financial year, we will be much more likely to succeed.

Vendor/Customer Relationships

People may have formed the impression that Japanese companies are hard, tough, and uncompromising. While there is clearly a degree of truth in this supposition, we can and will be very supportive of our suppliers in times of difficulty. Of course we recognise that problems occur from time to time, but we impress upon all our suppliers the need to let us know of any problems immediately they arise. If we know, we may be able to provide some assistance in resolving the matter. If we do not know, then we quite understandably assume that deliveries will be made in strict accordance with the schedule. It is no use telling us of a problem on the day of delivery: By then, it is far too late.

Everybody knows the philosophy of changing 5 percent of your suppliers every year, in an attempt to strengthen the market for procurement. I personally would not want to follow it. It is rather like making an example of the little boy who does something slightly wrong at school to frighten off the others. I do not believe in that philosophy, and I think in the long run, it will not do anything to enhance supplier relationships.

We like to develop long-term relationships with our suppliers. By doing so, we achieve a far greater understanding between the two companies, and it is important to recognise that this understanding must be at all levels, namely:

Managing Director	– Managing Director
Sales representative	– Buyer
Production controller and even	– Production controller
Lorry driver	– Goods receiver

We do not change suppliers lightly, and although we had to in a couple of instances, it was only after a long period of much effort to make the relationship work: Only after it is apparent that there is no future in the relationship will we take the final step of terminating it.

We also want to make our suppliers feel they are part of Toshiba as a family, and although we have a long way to go to achieve what has been done in Japan, we have made some progress. To foster this idea, we have held three supplier conferences. These conferences bring together our vendors and let them see our factory, reinforce our fundamental principles, and give an insight into the future. This also gives an opportunity for our suppliers to meet each other socially, and it is quite common for two suppliers who are in direct competition with one another to discuss their business not only as regards TCP but also in a broader context. In addition, these conferences provide an opportunity for some of our suppliers to voice their own views, and we invited three suppliers to talk at our last conference.

One of these demonstrated that his own business had substantially improved since becoming a supplier to TCP. In fact, he had at one time increased his own business so much that he almost came to us to say he could not supply our requirements as we grew. However, when we pointed out to him that his business had improved primarily as a result of our relationship, he rapidly backed away from this suggestion. Another made an impassioned plea for us to review our ordering policy so as to reduce the paperwork involved: a point we listened to with some sympathy. We have now changed from a job ordering policy to a stock number ordering system. This change is greatly appreciated by all our suppliers.

In Japan, our parent company holds suppliers' conferences twice a year, and in conjunction with them, they have formed a Toshiba suppliers cooperation group. This group meets regu-- larly three or four times a year and every two years organises a foreign tour. Toshiba will discuss any problems relating to their business with them, and together they will attempt to find a solution.

Just in Time

Our growth is such that in terms of colour television alone, we are five times as large as when we started. This has put considerable pressure on space, particularly in terms of storage capacity, both for finished goods and for purchased materials.

We do not have a major problem with finished goods since we generally manufacture what the market requires. However, we have considered the materials situation in some depth.

We now purchase small components by stock number and accept deliveries twice a month. Thus, for the first batch in any half month period, we will acquire the components only shortly before they are needed. For the last batch in the half month, these materials will be in the company just over two weeks before they are required.

Large components, such as picture tubes, cabinets, front bezels, and backs for example, are delivered on a batch basis, one to five days before production. Again, as a result of our expansion, this time is getting shorter, and stock levels are being reduced as a consequence.

Finally, for packing material such as cartons and expanded polystyrene cushions, our delivery lead time is now well under a day, and in many instances we receive two deliveries per day, thereby operating a true JIT concept.

In summary, our own concept of JIT is one in which we have first studied the components (and supply sources) and established a set of rules and methods that best suit our individual circumstances, such as:

1. Far East supply – Order exactly what is required on a monthly basis but recognise that because of shipping schedules, JIT cannot be operated.

2. Local small parts – Order exactly what is required on a monthly basis but with two deliveries per month. For small items, it is not practicable to adopt a true JIT approach.

3. Large parts – Where possible, we order by batch with a lead time of one to five days before production, approaching JIT.

4. Packing – Order by batch with delivery on the day of production or the day before for EPS -- two deliveries per day (JIT).

Much of the above is conditioned by vehicle load sizes. It is clearly practicable and most economic to deliver full loads, and as far as possible, we organise our deliveries this way. This does mean that in a number of cases JIT cannot be operated, although the lead time for the first use of the load is obviously reduced to a minimum.

Finally, in order to implement JIT, the manufacturer cannot operate in isolation. It is vital that suppliers are totally involved in the decision, that they realise their responsibilities, and appreciate the consequences of any failure to deliver to the required timescales, as any lateness can only have one effect: to stop production. And that is a cardinal sin in any manufacturing enterprise.

In conclusion, I do not want anybody to be under the misapprehension that we operate a very efficient JIT system. We do have some parts we consider just in time, and we have other parts that we do acquire in bulk. However, what we have always done from the outset is to order our exact requirements on a month-by-month-basis. So we do have a policy of not holding any stock except what is required for the following month. Now that helps us tremendously. Apart from the obvious avantages of a reduction in stockholding, one of the other big advantages is an enormous reduction in the amount of storage space you require.

If I can just amplify the significance of having local suppliers: To a degree, local suppliers are absolutely vital in many instances. We do not have many very local suppliers, but with those we do have, we find we are in a position to operate a much closer JIT concept in comparison with those that are some distance away. Generally speaking, the type of components with which you could operate that system are those components that are very specific to you and could be termed sub-contract items. Taking Mullard's point (see p. 47ff) in terms of proprietary electrical components, the economy of scale would not really be for a supplier to set up a plant close to each of his major customers. So you obviously have to accept that you are going to have your supplier some distance away.

In looking at experiences in our parent company in Japan, it is undoubtedly true they are very much closer to their suppliers. Even so, we have made tremendous attempts to get as close as we can to our suppliers.

But I still believe the parent company's stockholding (and their manufacturing volumes are much higher than ours) was considerably less than ours -- and they get daily deliveries even of small components from most of their suppliers, particularly such items as semi-conductors.

I also heard in Japan that with many companies operating JIT concepts, a major motorway accident brought practically the whole of Japan to a standstill, because all the components were on the road.

38

I think our company in the UK has been very fortunate not to have had any serious disruption of an accidental nature. The worst delay I can remember was when some Far East components were mislaid in a container at the docks. This disrupted us for two or three days. It meant we had to do a certain amount of schedule changing and caused some line stoppage. It also meant incurring an increased cost to air freight in some components to get over the situation as quickly as possible. The irony of that particular situation suddenly came to notice a year later, when the goods turned up. There was an enormous time lag, but we have not had any serious disruptions that have given us major cause for concern.

But our own experience has grown, and we have a four-months' lead time from planning to final manufacture: That is the current situation. When we started, we had a six-months' lead time, which we reduced in two stages to five months and then to four months. We are now looking to reduce that still further, but you cannot just suddenly implement JIT systems. They have to be worked through gradually, and I believe by doing it that way, you will make sure you do not suddenly find yourself in chaos. The other thing I would stress is that it is vital not to neglect the detailed planning activities. In fact, the shorter you decide to make your lead time and the closer you get to JIT, the more important is the planning to make sure you can achieve your aims.

Figure 1

LEAD TIME SCHEDULE

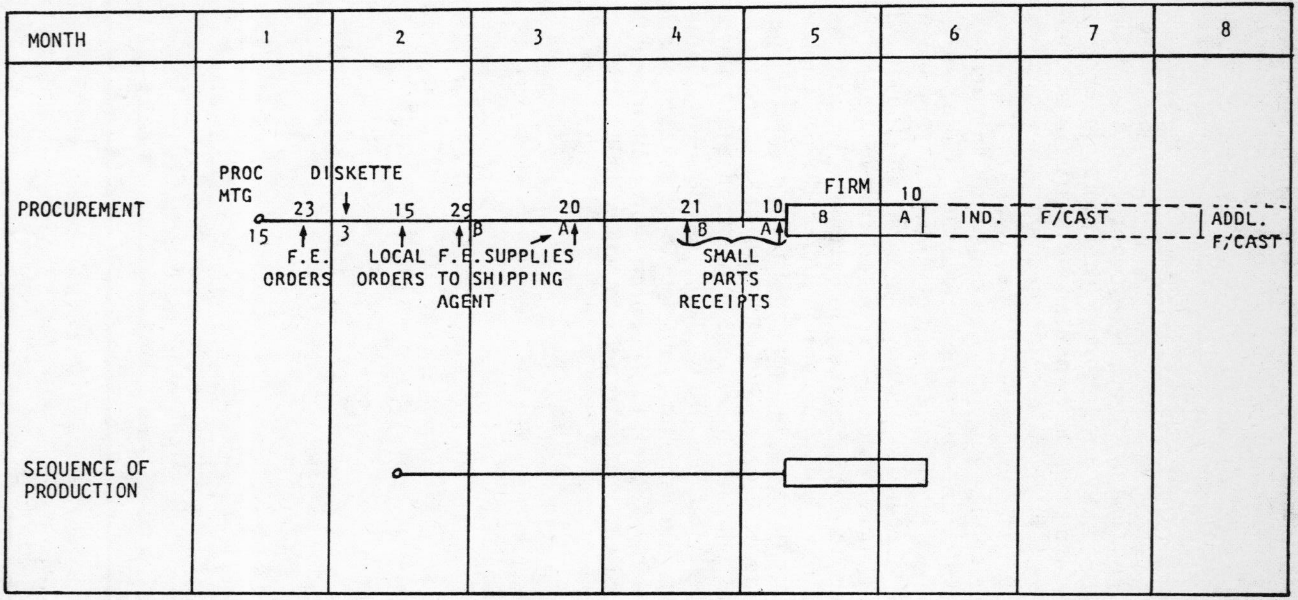

MONTH	1	2	3	4	5	6	7	8

PROCUREMENT

PROC MTG — DISKETTE

15 — 23 — 3 — 15 — 29 — 20 — 21 — 10 — 10

F.E. ORDERS — LOCAL F.E. SUPPLIES ORDERS TO SHIPPING AGENT

B — A

FIRM

B — A

IND. — F/CAST — ADDL. F/CAST

SMALL PARTS RECEIPTS

SEQUENCE OF PRODUCTION

Figure 2

INITIAL SYSTEM

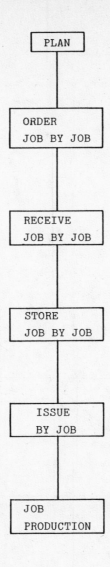

Figure 3

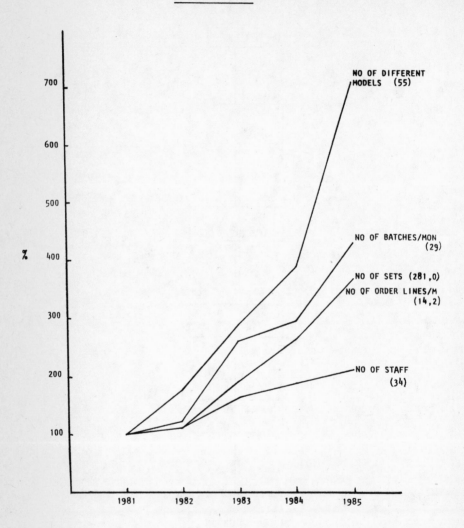

VOLUME GROWTH

Figure 4

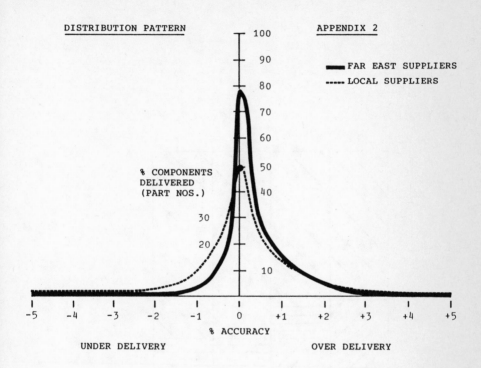

DISTRIBUTION PATTERN

APPENDIX 2

FAR EAST SUPPLIERS
LOCAL SUPPLIERS

% COMPONENTS
DELIVERED
(PART NOS.)

% ACCURACY

UNDER DELIVERY

OVER DELIVERY

43

Figure 5

ORIGIN OF PURCHASES (1)

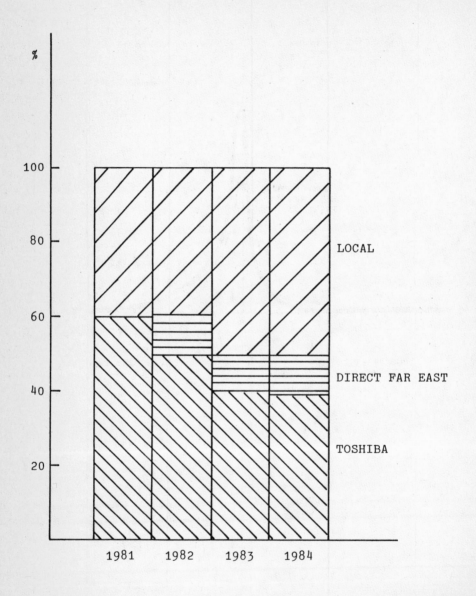

Figure 6

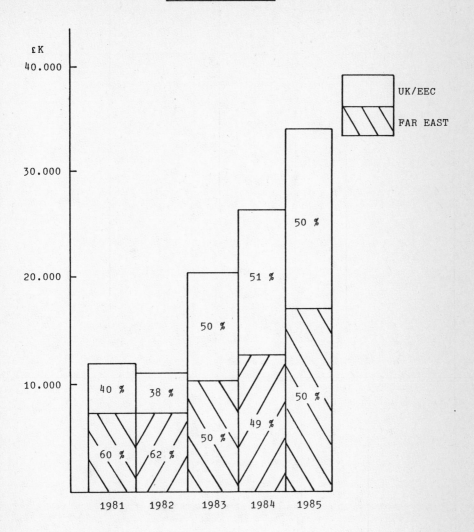

CTV MATERIAL COSTS 1981 - 1984

FAR EAST V UK/EEC

EXPERIENCES OF A BRITISH FIRM SUPPLYING COMPONENTS TO JAPANESE COMPANIES

Duncan Edwards
Marketing Manager
Consumer Division
Mullard Ltd.

Mullard embraces the UK activities of the Philips worldwide electronics component operation including marketing, R & D, and production. Mullard sales in the UK are in excess of 300 million pounds, supplying components for consumer equipment manufacturers (radio, audio, TV, domestic appliances) and for professional equipment manufacturers (computers, telecommunications, instrumentation, process control, defence equipment, etc). Over 50 percent of our output is exported from the UK, and we account for 15 percent of the total exports of the whole UK electronic components industry. Investment in the UK amounts to 250 million pounds. Seven thousand people are employed at eight major factory complexes, plus the head office and our research and development facilities.

A significant proportion of Mullard's turnover has been with TV and audio equipment manufacturers. Mullard are the largest supplier to this sector of the market, supplying a wide range of products from the TV tube to small electronic components. In recent years mechanical components such as printed circuit boards and plastic cabinets have been added.

The TV-setmaker market has seen a substantial restructuring during the last 15 years. At the beginning of the 1970s, Sony established a TV-set manufacturing unit in South Wales, and over the intervening period, some six major TV plants have been established by other Japanese setmakers. Many of the traditional UK-owned companies have subsequently left this business sector or indeed have stopped trading altogether.

Mullard sees the TV component market as strategic, particularly bearing in mind the importance of manufacturing colour picture tubes -- a key Mullard activity. It was therefore necessary to establish business relations with the new Japanese operations in the UK to maintain our market position. Levels of penetration into new designs have steadily increased. Sales have expanded above the rate of expansion of the total market, with Japanese setmakers becoming major customers for the Mullard company. During this period, set production in the UK has doubled to three million units per annum, and Japanese companies will provide 46 percent of the total manufactured this year. During the restructuring of the industry, Mullard had to learn how to become effective suppliers to Japanese customers, which has involved a considerable reassessment of the way we looked at our business.

The first task we had to address was learning how to communicate effectively with colleagues from a different culture and background, but there is no such thing as a standard Japanese company. As in Europe, each company has its own style; so that Matsushita will be different from Toshiba and they from Sony, etc. Obviously, a companys' style reflects the skills of its founder members and the way the company has expanded subsequently. Some styles push decisions to the lowest level; others will be relatively autocratic.

The problems of communicating with another person who has grown up in a different culture and background are difficult to anticipate. It is quite troublesome across Europe, as somebody who works in a multinational company can testify. The way an individual looks at or studies a problem is very dependent on the traditions of the country in which he lives.

Some Aspects of the Initial Problems Experienced

The most obvious obstacle is <u>language</u>. We expected Japanese colleagues to hold discussions in English: a language that is highly complex and has many shades of meaning. A person from Great Britain delights in expressing the same thought or statement by using a different set of words if he is asked to repeat something. This caused great confusion initially because our Japanese counterparts saw it as a totally different statement requiring a rethink before answering. Discipline in language use is an urgent priority so there is no misunderstanding.

The mechanism for <u>reaching a decision</u> and then implementing an agreed action is fundamentally different between European and Japanese cultures. Achieving consensus on an agreed plan before activating it is a way of life in Japanese business, but in the UK, a manager will often act on his own initiative before a team consensus is reached. In Japan, reaching a decision can take a long time, but once achieved, an organisation can implement it quickly because everybody understands what is required. In the UK, implementation may be slower because the people involved are still learning what is required during the implementation of the project.

Another cultural difference is the <u>attitude to work</u>. In Japan, a man's work function is viewed as the dominant factor in his life, to which everything else is subordinated. In the UK, it is viewed as one of a number of aspects that are important. What is normal in the UK situation, where a factory closes for a three week-vacation, is unknown in Japan. A Japanese worker finds it difficult to be away more than a few days because of the stress his absence causes to his fellow workers.

In reaching a position of trust between our potential customer and ourselves, we learnt that the <u>role of supplier</u> is not viewed as that of a partner at the design phase. Mullard, with experience of chassis designs in Europe, was able to offer total design concepts, but this was an unusual role for Japanese component suppliers.

This proved to be a major problem -- the design authority was based 11,000 miles away in the Japanese parent company. Japanese setmakers have found it difficult for many reasons to establish TV design centres in the UK, and the rate of progress has been disappointingly slow.

So having outlined some of the problems, it is necessary to explore them in more detail before saying how we at Mullard have tried to deal with them. As already indicated, the solution was only reached by discussion with our Japanese colleagues once a good relationship had been established. But it can take several years before this position is reached.

A good <u>relationship between individuals</u> is most important in Japanese business, and every effort is made to encourage fellow workers. So the use of the word "no" is almost unknown. To say "yes, but," pointing out the great difficulties involved, is viewed as a better way of handling a situation. But it does mean that you have to become very sensitive to the way the word "yes" is being used.

Great sensitivity is also required in <u>presenting requests for action</u>. Often they will appear as suggestions not forcefully put, but nevertheless they require immediate action or disappointment will result, causing friction and misunderstanding.

Personal commitment on the part of the individuals concerned in customer contact is one essential requirement for a successful business relationship, but it is no good if the front man is not backed up by an equally dedicated support team within the supply company.

Constant improvement in everything achieved is another requirement for a Japanese supplier: Both products and service must show continual improvement. Delivery to an agreed quality level (AQL) is not acceptable. Any failure, for whatever reason must be analysed, the reason for it defined, and an action or countermeasure must be found and implemented so that wherever practical, a possible repeat of the failure is eliminated for the future.

In Mullard's case this attitude was the most difficult to introduce, but as can be seen from the results, it has now been successfully adopted by our organisation.

So what has been the Mullard programme to meet the requirements of doing business with Japanese customers? It has included the following measures:

- A Japanese business culture course:

    This was a three-day seminar in which lecturers, some of them Japanese nationals, helped us explore the differences between our cultures, Japanese business practices, and the role of a supplier. Every member of the company's senior management, including the financial and production managers, attended with account managers, salesmen and the sales desk, and logistics managers. During the three days, we attempted to gain a common knowledge across the company of the likely customer requirements and the response required from Mullard.

- We invited Dr. J.M. Juran, the guru of quality thinking, to lead us for a weekend course on the nature of quality and the major tools for measuring it.

- In order to increase awareness throughout the Mullard organisation that quality was the result of continuous improvement, all middle and senior management attended a two-day seminar aimed at encouraging commitment to the attitude that any failure at all was not good enough. The philosophy was that improvement depended on each individual's contribution and that each individual had to be committed to playing his part.

- Junior members of the company were given a one-day course, and quality improvement teams were established across the company in the commercial operation and in the factories.

- Measurements were taken of five company performance indicators. Targets were established and progress towards these targets monitored. The first areas addressed related to deliveries, stock, cash flow, and returns. Departments were asked to contribute improvement programmes covering their particular responsibility.

- After two years, a review for each site was held involving every member of the company so that they were all aware of progress made and the targets for the next period.

- Each individual and section has now been asked to submit its own particular improvement plan.

This programme has now been integrated into a Philips company-wide activity. It has required a major change in company philosophy to achieve the results targeted, but progress has been steady: It cannot be achieved overnight, because of the need to gain understanding and commitment from each individual in the company.

In parallel with these internal activities, Mullard has been providing customer support in a number of areas, such as:

- visits to the customer in the UK by directors, senior managers, and account executives to develop personal relationships at all levels, to present the resources of the company, and to underline our commitment to the business;

- visits to Japan by the managing director and the commercial director to establish relationships with heads of agreement with counterparts in Japan. These visits now take place around every twelve months as appropriate;

- visits to Japan by senior experts able to assist in chassis design and any specific problems. This was an attempt to reach and meet the design authorities in Japan. These visits are planned every four to six months to coincide with a new chassis cycle as appropriate;

- a liaison office has been established in Tokyo staffed by Japanese colleagues, whose role it is to maintain a Mullard presence in Japan on a day-to-day basis. They monitor progress and make sure problems are followed up.

  Information can be sent either through the UK operation to the parent company in Japan or through Nihon Philips, depending on where the query emanates from.

Strict discipline was self-imposed so that promises were accurate and misunderstanding could be avoided between companies separated by 11,000 miles. Response times must be good and failure to meet a promised deadline causes major difficulties.

Everything has been targeted to encourage a zero-defect approach in our customer relationships. As indicated earlier, there has been considerable success even though we are well aware that there is room for further improvement.

In trying to find an independent measure, use has been made of the figures produced by the National Economic Development Office (NEDO) in the UK. NEDO provides a forum for consultation between government, management, and unions, and in a recent paper it documented the improvement in reliability for various aspects of the TV manufacturing industry.

The results for high-value components, such as integrated circuits (ICs), tuners, CRTs, etc. for the period 1977–82 show a 25-fold improvement in ICs (37,000 to 1,500) and a 37-fold improvement in tuners. CRTs have now reached 6000/million, which is a low figure because it includes faults found during adjustment and mounting. Even more dramatic progress has been made in small components. The overall result has been that the reliability of TV sets has improved dramatically during the period 1977–82 and, of course, further progress has been and still is being made towards zero defects. The result is that UK-produced sets now match the reliability levels achieved by sets manufactured in Japan. The correlation between long-term reliability and production quality can be seen in Figure 1 on page 55.

The emphasis so far has been on building an effective relationship with Japanese customers.

Where Does JIT Fit into This?

At its best, JIT involves the optimum matching of the resources of the supplier to that of his customer, so that <u>functions are not duplicated</u> between the two. As Peter Bayliss of Toshiba has stated, a prerequisite for introducing JIT operations is a good delivered quality of known conformance shipped on an agreed day. This position has now been reached with many customers, and we are now turning our attention to JIT principles for the next phase of our improvement plans. We see JIT involving a large number of disciplines beyond logistics: Quality, economic batch size, and flexibility are some of the elements.

Despite all the accurate planning, one has to maintain flexible capacities. However I would say Japanese companies in general give you a much clearer indication of the likely shape of the business. In the sense that they are going to try, for instance, to double our business, they are much more accurate in that prediction. They do not necessarily say what they are going to make, but they will say, we think you will need the resources to take business up to double. We have to take a judgement on what that is. Other Japanese companies do not do that. In fact, we have had one recently that has just moved up by factor of a third within a period of six weeks.

It may be slightly contentious, but the use of JIT as it is often understood in the case of the car industry is limited to the bulky high-value items required for the electronic equipment business. It makes sense to move CRTs and plastic cabinets directly between the supply factory and the equipment maker, but it is not yet clear what the best method of supply is for small components of low value. The volume of components consumed is large, but it involves a complexity of types, each of relatively low volume.

It would be interesting to know how other companies are addressing this problem of avoiding holding large stocks of components and yet maintaining flexibility in their production programme.

Another point I would like to explore is geographically positioning the supplier close to the user. In my industry, some of the investments you need, for instance in television tubes, are enormous. It does not make sense to have very many units making TV tubes to supply the European industry, for instance. The second thing is that we are in a time when technology, certainly in electronics, is changing, and there are machines around now that are fairly flexible. These can produce large amounts of equipment using a technical surface mounting. I envisage that there could be a restructuring of the industry; perhaps the surface mounting plant could make TVs in the morning, radios in the afternoon, and baby alarms in the evening. That would be a useful way of using the capital.

Since interruptions are an inevitable part of the business scene, a company operating a JIT system must take active precautions. The way we try to tackle them is really twofold. One approach is for the catastrophic failure of a plant, or a fire, or something like that. We try to handle that by having at least two plants around the world that make any one technology.

So at no time are we 100 percent dependent on one plant for one product. The other approach covers the situation if the supply chain is wrong for any reason. We would try to shorten it either by flying or by putting people in cars to get things. We had a case last year where we were in the process of doubling the capacity of a CRT plant in Austria at a time when the market was going up and up and when Japanese resources were being switched to China for various reasons. Not only were we short of capacity, but so were our colleagues in Japan. The more you become conversant with the situation, the more you know how to manoeuvre supply chains. But basically this is not something that happens very often: perhaps twice a year as far as I am concerned. And it is dealt with either by putting in from a different plant or by shortening the supply chain.

Figure 1

RELATIONSHIP BETWEEN RELIABILITY AND PRODUCTION QUALITY, UK-MADE COLOUR TV SETS 1977/82

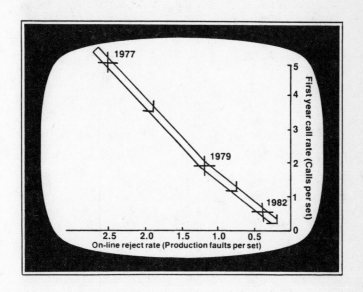

A BRITISH TRADE UNION VIEW OF ORGANISATION AND JIT IN THE UK

Ken Cure
Executive Council Member
Amalgamated Engineering Union

Every generation has dealt with new technology. Indeed one of the earlier industrial disputes in Great Britain concerned the introduction of new technology although this dispute does not get a great deal of pubicity these days. This was the strike of the South London water carriers against the introduction of piped water in the capital.

People Are the Key to JIT

It is not too clear whether JIT comes under the heading of technology, management, philosophy, or something else. But one thing is certain: People are the key. People are involved throughout -- and this applies to any new approach. People are involved right through the whole of any technological advance -- in the same way that the word "Blackpool" goes right through the stick of peppermint rock known as "Blackpool rock." Any technological advance is people-driven and is carried through by people. In that respect we have to make a better use of people. We have to make things work better and more effectively because we live in a competitive world.

That does not mean we have to think only of JIT as representing another danger, particularly in the UK. JIT is a logistical exercise. It means delivering parts and components to where they are needed to make a greater whole. In this respect, management has to be very much more expert than it has been previously: The unions are waking up to an alibi that has been used in the past to explain away lost production. If one looks at industrial disputes in the motor industry, for example, there have been many occasions when the excuse of a dispute has been used as a convenient smokescreen. In actual fact, what has very often happened is that there has been a failure in the supply chain, for which an industrial dispute is a handy alibi. In those circumstances, it is not surprising that people employed in industry say the way things are done is ridiculous on that kind of time scale and using those methods.

One aspect of JIT is that it is all about talking together and consulting people. To use the earlier illustration of Blackpool rock again, there was a dispute at the manufacturers. As a result, one man was sacked but was not off the premises immediately, and he made his feelings known by making the "rock" with a rude word going all the way through it, which the management, of course, could not sell.

JIT Results in Quality Improvement

As a trade union, and this applies particularly to my own union but also to others, we are just as committed to improvements in manufacturing industry as the management is. After all, there are jobs at stake, and manning levels have decreased rather sharply in recent years. In the final analysis, JIT results in quality improvement (or at least it is supposed to) increased productivity, lower costs, on-time deliveries, and improved motivation or morale, which comes back to the people aspect again.

Some time ago, John Egan, managing director of Jaguar cars, was talking about quality. One of the things he said was, "Nobody ever saw an alternator broken down on the hard shoulder of a road. Nobody ever saw, or took much notice, of a deflated tyre on the hard shoulder. What they saw was a Jaguar;" so components quite obviously are an integral part of the whole. In fact, bad components are twice as bad as short components because bad components have to be reworked and replaced. Short components get used up, but if you have delivered something that is faulty, you not only have to apologise, you have to reverse the impression given by the failure.

Look at the average body shell of a motor car. It consists of somewhere in excess of 2,000 pieces. Now, that is a logistical exercise that defies imagination, but nevertheless it has to be achieved. And over a long period of time in that industry I have experienced incidents where first-line supervision has said, "Never mind about whether it's right, get it out. We can put it right later."

Now, that is a long way away from a JIT concept, but it has existed for a long time, not only in Great Britain, I suspect, but in other places as well. So you really do have to make sure that the actual lesson itself is learnt at the interface. First-line supervision is responsible, and it is useless to preach to the shop floor about loss of quality if your management has not got it right. It is at least a two times equation: If you have a shortage, you have not got a product. If you have a wrong product, you have to put it right, so that you are working it twice: That is what I mean by two times. But more than that, you have to retrieve your reputation, which has been lost in the process -- and that is a long process. It is not just your reputation as an employer, it is all the people who work for you because they rely on the word that comes through it. I have even seen instances fairly recently when I have been round the stores in a company and seen the spares for a machine that was sold ten years ago or scrapped. You know, that is a managerial responsibility; just the same as training is.

Now, I spoke about training earlier, and this is where you come to getting it through to people. The present best practice in manufacturing industry in Europe (especially in the UK, which I know more about than about the rest of Europe) is about 2 1/2 percent of all working time: That is an average of all working time spent in training and retraining. With the new technologies that are developing, and JIT is one of those tools, you are going to need somewhere around 15 percent of all working time on average spent in training and retraining. Otherwise you will not get the concepts of JIT across. They will not get across to people or into the system, and it is useless just expecting people to manage with better expertise. Every craftsman that I know of these days is a systems manager, and I am sure my colleagues from IG Metall in Germany recognise precisely what I am talking about. If you are dealing with chip-controlled machinery, you are managing a system. And you have to understand what that system is before you start to manage it. This is something that indeed is again within the concept of management's understanding what it is all about before they even try to get it across to their own work people. It is something that is people-driven, is driven down as well as upwards. And you can remember that.

Consultation and Team Work

To make all this happen, a great deal of time and attention must be devoted to assessing the people implications. At the advanced manufacturing systems group with which I have been associated at the National Economic Development Office

(NEDO), for example, one thing that came out very clearly in all the discussions and case studies we dealt with in producing our report was that you could not begin to plan too soon, and you could not involve the people in the planning too soon. It had to be from the conceptual stage, not just, "By the way, fellows, we are bringing a new Cincinnati machine centre into the shop tomorrow, and we expect it to be running by the end of the week" -- because it does not work that way. It has to be planned in such a way that it fits in with what you intend to do, and as someone else mentioned earlier, it will not work if people do not want it to. You have to carry them with you. Everyone, and that means everyone, has to be made aware of the existing situation and the implications, and then they must be given a clear understanding of the techniques needed to bring about the required changes.

Ownership Must Be Exercised by Everybody

JIT is not the only discipline within this area. Ownership must be exercised by everybody. Now this does not necessarily refer to cooperatives, although one could discuss them. The point is that the time used in any enterprise belongs to everyone in that enterprise, not just the management -- and that means communication, education, and teamwork. It means that the trade unions as well as the management are all part of the team. Let us be adventurous. Let us speak to each other. Let us listen to each other, which is even more important. Let us think about progressive employment policies, constructive reward systems, different patterns of working, including shorter hours, or perhaps fewer hours.

Sabbaticals in the plants and continuing education have already been mentioned in the Jaguar case, and it is worth taking a look in their plants.

Above all, it is important to make sure that the interface between first line supervision and the people on the shop floor is good and that people know the company's plans and the reasons for those plans.

It is time for management to show us they want teamwork. Or do they? You will not get compliance unless people are involved. Get the people out of the boxes; let them cross boundaries. We have been saying it for years, but what has actually happened is that there has been a requirement to try to reduce the skills people use in the workplace into smaller and smaller capsules -- and then you are surprised that you get poor results.

In Britain and on the Continent, the word flexibility has two different meanings. At the moment in the United Kingdom, it means flexibility between trades rather than flexibility of time.

We are having talks about economy with our employers' federation, and here the point I would like to stress is that it is a question of 15 percent labour costs and 45 percent inventory. If you reduce the labour cost or the labour content, the money level or whatever you call it, you are reducing your flexibility. You are reducing your flexibility with far greater certainty than you are reducing it by not having people able to work all the days on the calendar and until you reach the idea that you are getting far more profitability for the company by reducing inventory: By recognizing that if you do not sell it, you should not make it. That is a very basic concept and one that is really behind JIT. Now, the burden of the workforce is to produce as effectively, as efficiently, and as humanely as possible to produce what you require for the marketplace.

That means, as a precondition, that they need a degree of confidence, that they have a future. That is also inherent in the Japanese scene, but I have not seen it applied anywhere in Europe yet. There are two sides to every bargain, and if you want the flexibility you are talking about, you have to give the guarantees of future employment that are the other side of it. You should not put an unnecessary burden on the workforce that they cannot sustain because it is not their burden.

Educating for a Change in Attitudes

Let people develop their potential. We have to educate for a change in attitudes. The Amalgamated Engineering Union is doing its part in what may be considered a fairly imaginative and responsible way. There are two programmes, one of which is called "Engineers 2000." It provides technical training and practical training for which employers are expected to pay because they get an added value from our members, their employees. This programme does not deal fully with skills, but it at least provides very solid signposts for the way people can go. We provide, for example, the practical aspects of the Open University course on robotics in manufacturing and do the actual hands-on part of that course. Our members, if they want to, can follow the rest of the course afterwards. It is recognised as part of the course. In what we call "Geartrain," we are developing lessons in effective industrial relations.

At the end of 1986, part of that was taken up with pilot courses that compared the alternatives as well as JIT philosophies. We will have problems not only now but also in the future unless our people understand what is behind JIT.

Cost of Inventory

One thing I think you have to remember when looking at JIT philosophies, in particular in manufacturing, is the cost of inventory. In the Financial Times recently, there was a report showing that in British companies, a very large amount of money was quite unnecessarily tied up in systems, which in contrast to just in time, might perhaps be called "just in case." As far as the UK is concerned, the Diesels Economic Development Committee, of which I am chairman, recently commissioned a study on costs in that particular sector, which are indicative of the rest of manufacturing. About 15 percent of the total costs of enterprises in the manufacturing sector in the United Kingdom consists of labour costs, including the salaries of directors. 15 percent! But about 45 percent is the cost of inventory. So why take it out on us?

We have to develop new styles, and we have to be imaginative and adopt new procedures. Our people look to maintenance of employment, and that is one of the aspects of the Japanese experience we find rather more attractive. Our people want to feel useful. They want to be effective. They want to be part of a total team. They want to be able to feel some pride in the product they are producing. But to do all that, management need to remember that the people who work for them are people. To reiterate, JIT is about skills, attitudes, styles, procedures, and above all consultation.

APPLICATION OF JUST-IN-TIME MANUFACTURING TECHNIQUES IN THE UNITED KINGDOM*

C.A. Voss and S.J. Robinson
University of Warwick

Background

The Japanese have been highly successful in designing, manufacturing, and marketing a wide range of products that have penetrated markets worldwide. The extent of this success has become so noticeable that there is now considerable interest in Japanese manufacturing techniques, of which those classified as "Just-in-Time" (JIT) techniques are perhaps the most widely discussed. There is currently little knowledge of the extent to which these techniques have been put into practice in the United Kingdom. This chapter reports the results of a survey of UK industry designed to develop knowledge of JIT application. In this research, a broad-based view of JIT is taken, encompassing JIT in manufacturing, purchasing, and supply (see Appendix I).

Objectives and Sample

The objective of the survey was to determine the extent of knowledge of the JIT concept in the manufacturing industries in the United Kingdom, together with the level of application of various JIT techniques and, if possible, to obtain some indication of the ranking of any perceived benefits that may have been derived from their implementation. The scope of the survey is limited to this assessment of JIT: Thus, it is not proposed to judge whether the implementation of JIT would be either appropriate or effective in any particular manufacturing environment.

---
* Our thanks to Keith Howard, Editor, I.J.O.P.M., MCB University Press, Bradford, England, for the permission to extract this chapter from the "International Journal of Operations and Production Management"

The sample comprised addressees obtained from two sources: A list of 300 company representatives who had expressed interest in attending conferences in manufacturing technology and a random selection of 200 company representatives taken from a list of UK manufacturers. The aggregate list was checked to ensure that no duplication of companies occurred. A response of 123 good replies (25 percent) was obtained. The survey was backed up with interviews in five companies known to be adopting JIT.

Table I: Application of JIT in Manufacturing

Implementing or Planning to Implement Some Aspects of JIT (%; n = 132)

Yes	57
No	43

Table II: JIT Effort in UK Companies

Formal Programme for Investigation and Implementation of JIT (%; n = 132)

Yes	16
No	84

Table III: Nature of JIT Effort

Nature of Effort	% of Total	of Those Implementing Planning to Implement
Nil	50	17
Experimental	14	20
Ad hoc, modification of existing systems	26	46
Major JIT programme	10	17
	n = 132 (100%)	n=70 (57% of total)

Survey Results

Overall Consideration of JIT
The survey found that 57 percent of the sample either plan to implement, are implementing now, or have implemented aspects of JIT (see Table I). It was also found that 59 percent felt they had either some or a good understanding of JIT. This is indicative of the level of interest in particular JIT techniques, as well as a high level of interest in the concept of JIT.

Nature of the JIT Effort
Despite the widespread interest shown, only 16 percent of the sample stated they had a formal programme for the investigation or implementation of JIT (see Table II).
Companies were also asked about the nature of their JIT effort. The results are shown in Table III. These responses suggest that, despite the strong interest being shown in JIT, only a small percentage of companies have a major commitment to JIT at the moment. Only 10 percent of the respondents have a major JIT programme, although 40 percent were conducting experiments, ad hoc programmes, or modifications of existing systems.

JIT Techniques Used
Companies were asked to indicate, from a given list, which aspects or techniques they were using. Table IV lists the results both in intention and in terms of techniques already implemented.

Table IV. Ranking of Aspects of JIT Chosen for Implementation by Companies in "JIT-Considered" Sector

Aspect	Aspects implemented as % of Total Sample	Aspects Planned, Implementing, or Implemented as % of Those Implementing or Planning to Implement JIT
1. Flexible workforce	30.1	80.0
2. WIP reduction	18.7	67.1
3. Product simplification	16.2	60.0
4. Preventive maintenance	11.4	60.0
5. Statistical process control	13.8	58.6
6. Set-up time reduction	16.2	54.3
7. Continuous improvement	11.4	54.3
8. JIT purchasing	15.4	51.4
9. Work team quality control	11.4	50.0
10. Standard containers	15.4	44.3
11. Modules or cells	11.4	44.3
12. Zero defects	3.2	34.3
13. Mixed modelling	8.9	31.4
14. Smoothed line build rate	9.7	25.7
15. Parallel lines	10.6	22.9
16. U-shaped lines	9.7	22.9
17. Kanban	4.1	11.4
	n=123	n=70 (56,9%)

There are a number of items that stand out from this table. First, the high proportion of companies who are looking for and have implemented flexibility in the workforce. Second, the small number of companies that have implemented what might be termed the core techniques of JIT, such as cellular manufacturing, statistical process control, U-shaped lines, schedule smoothing, and Kanban. Kanban, the best known of all the techniques, comes next to bottom both in actual implementation and bottom in planned implementation. It would seem, from the data in this table, that less than 15 percent of companies in the sample have implemented a wide set of JIT approaches or techniques. Even the majority of the companies planning or implementing JIT have considered only a partial approach.

It may also be significant that the measures that rank at the upper levels are generally easier to implement than some of the measures at the lower levels, which entail a commitment to the JIT concept. Clearly, the introduction of measures such as Kanban or U-shaped lines would require substantial rearrangement of the production facilities: The comments made by companies interviewed support this and lead to the conclusion that many companies are implementing indiviudal aspects of JIT rather than the whole concept. However, this should not be constructed as a negative statement; clearly, there is considerable interest in the subject, and the very essence of JIT is to continuously strive for improvement.

Table V:Benefits from JIT: Techniques of JIT Leading to Significant Benefits (Ranked by % Claiming Significant Benefit from that Technique)

Ranking	Technique
1	Zero defects
2	WIP reduction
3	Kanban
4	JIT purchasing
5	U-shaped lines
6	Work team quality control
7	Modules and cells
8	Set-up time reduction
9	Flexible workforce
10	Parallel lines

The Effectiveness of JIT Techniques

For each technique, companies that had implemented that technique were asked to rate, on a four point scale, the benefits achieved. Table V lists the techniques in rank order of the percentage of users reporting significant benefits from that technique. The ranking is based on a very small proportion of the sample: For example, all four companies that stated they had implemented zero defects claimed that the resulting benefits were significant, hence the highest ranking. However, despite this limitation in sample size, the ranking listed in Table V suggests that greater benefits have been achieved by techniques ranked at the lower end of Table IV. This indicates that "core techniques" of JIT produce greater benefit than the more frequently used and more easily implemented techniques.

Benefits from JIT

Companies who had implemented JIT were given a list of possible benefits and were asked to rank them. The results are shown in Table VI. Reduction in WIP and increased flexibility were consistently ranked as the major benefits derived from JIT, though a number of companies rated increased quality as the foremost benefit to be gained from the introduction of JIT.

Industry and Related Variables

Data were collected from the companies on these variables to identify possible differences between adopters and non-adopters of JIT. (Adopters were defined as those companies that had planned to or were implementing one or more aspects of JIT.) The results are shown in Table VII.

Table VI: Ranking of Major Benefits by Respondents Who Had Implemented JIT

Ranking	Benefit
1	WIP reduction
2	Increased flexibility
3	Raw materials/parts reduction
4	Increased quality
5	Increased productivity
6	Reduced space requirements
7	Lower overheads

Table VII: Industry Sector and Company Size

	% of Sample	% Adopting JIT
Sector:		
Electronics & instruments	13.8	58.8
Chemicals	4.9	50.0
Machinery	17.1	66.7
Electromechanical assembly	29.3	69.4
Consumer non-durables	11.4	28.6
Other	23.6	48.3
	100.0	
Company Size:		
Large	19.5	58.3
Medium	51.2	61.9
Small	29.3	47.2
	100.0	

It was found that the electromechanical, machinery, and electronics sectors formed a significant proportion of the sample. Also, the data in Table VII suggest that more consideration has been given to the implementation of JIT in these sectors than in the chemicals, consumer non-durables, and other sectors. The data also suggest that marginally more attention has been paid to JIT in the medium-sized companies than in the case of the large companies; small companies ranking lowest in this respect. However, the problems of classifying companies by size must be remembered when considering this result.

Nationality of Parent Company
A comparison was made between British companies and those owned by overseas parents to see whether there was any relative difference in their level of consideration of JIT measures. It was found that 55 percent of UK-owned companies compared with 62 percent for overseas companies were JIT adopters. This suggests that there was marginal but not significantly higher interest where there were overseas parent companies.

Summary

The survey has indicated a mixed view of the acceptance of JIT in the UK. On the one hand, it has shown both a high level of awareness and understanding and a large number of companies implementing or intending to implement some aspects of JIT. Even allowing for the bias inherent in the nature of the sample and for the possibility that those interested in JIT were more likely to respond, the general awareness of JIT is now high. This is not surprising given the frequent reports of major benefits achieved. The survey found that reduction in WIP and increased flexibility were consistently perceived as the aspects of JIT from which most benefit was derived.

However, when we review the findings on actual implementation, the picture is less favourable. First, few companies are actually making a serious attempt to implement JIT. Where they are implementing JIT, many companies are implementing just a subset of JIT, the data suggest that companies are focusing on easy-to-implement techniques rather than those giving the greatest benefits. This gives cause for concern. First, JIT is a holistic approach and many companies seem to be neglecting techniques or approaches that are central to JIT. Second, techniques or approaches yielding the highest benefits were those least frequently adopted.

Thus, though overall level of interest in JIT is considerable, many companies would seem to be implementing various aspects of JIT on an ad hoc basis while few are applying JIT techniques as part of a planned and integrated manufacturing policy. If UK companies are to move beyond interest into widespread and successful use of JIT, a more complete and committed approach is required than is demonstrated in this survey.

INTEGRATING JIT INTO A TOTAL PRODUCTION AND MARKETING SYSTEM

Hans Peter Stihl
Andreas Stihl KG

Severer competition worldwide demands quick reaction to changes in the market.

The application of JIT in an industrial firm means bringing the delivery capability of the whole firm into line with the requirements of the market on a just-in-time basis. Such a demand can easily be formulated as a corporate objective, but implementing it in the daily running of the firm is hard work. Anyone who consistently steers towards this goal will at first experience setbacks, rather than being spoiled by success; but anyone who wants to be competitive in world markets has no other choice. Ever sharper competition forces us to produce and to deliver in a more market-oriented and order-oriented way. As a manufacturer selling products to 130 different countries, 90 percent of whose turnover in 1985 was in export markets, I know what that means; but I also know that JIT systems can help me.

World competition has intensified due above all to innovations that have made microelectronics possible, and the application of this new technology is coming to be the equivalent of an industrial revolution. One consequence, and an altogether positive one, is that the different markets in the world are moving closer and closer together and becoming more transparent. Thanks to the application of microelectronics to communications technology, competitors throughout the entire world are pressing each other ever harder.

It is barely forty years since medium-sized business began to acquire the telex. Previously, anything urgent had been dealt with over the telephone or in a telegram, and larger objects could only be "transported" by post. Postal deliveries to other continents took weeks and months. But today we can transmit anything in fractions of a second by facsimile and other modern methods. The long-distance transmission of data is everywhere on the march, and to exaggerate a little, one can say that as far as communication is concerned, we are in the process of abolishing the time factor. This is not something that can remain without implications for the business world, and the introduction of JIT systems at company level is one consequence and one answer to the problem.

The Need for an Integrated Overall Logistic Concept

JIT cannot simply be introduced into a firm overnight. It assumes, for example, the existence of a new company infrastructure, and this takes years to attain. If JIT is to be really effective, it cannot be restricted to certain parts of the enterprise, and it is well known that the speed of the slowest vessel determines the progress of the convoy. What is needed is rather an overall logistic concept that takes full account of technical developments in process technology and process control in relation to procurement and the purchase of materials, development and construction, production and distribution -- and this concept can only be implemented on a long-term basis and put in place step by step.

The advantages of such a uniform concept integrated into a JIT system for the whole of the enterprise are above all evident in the following:

1. faster implementation and reaction times both in the marketplace and in the company and therefore greater flexibility,

2. faster flow of materials and assembly and therefore less need for buffer stocks,

3. lower warehousing costs and the freeing of funds tied up in floating capital thanks to reductions in inventory.

Naturally, an integrated logistic concept does not appear on a tabula rasa but, normally, out of an evolutionary process within the firm. This is what happened in the case of my own firm in Waiblingen which, with its electric saws and range of power motors, is among the mass producers. In our case, the first "pieces of the mosaic" developed in separate areas of the enterprise, and it was not possible to identify logistic connections between them.

Such an impetus came, for instance, from the distribution side. The fact is that we kept having unpleasant surprises in our various markets, and the more numerous the countries that we exported to became, the more difficult it was to obtain a clear view of our sales situation. It was difficult for us to oversee the stock position and sales results in the whole worldwide network of foreign importers, wholesalers, and retailers. Neither at the right time nor in sufficient depth did we get information about market conditions, market fluctuation, and changes in customers' attitudes, or about the activities of competitors. There, uncertainty factors led to a reduced ability to react appropriately. In addition, the lines of supply from our sole production site in Waiblingen to many of our most important export markets were too long, and it often happened that we worked overtime at full capacity when the markets had long ceased to take any more products. This was simply because we did not have the right market information at the right time. Under these circumstances we implemented two important strategic decisions:

1. We strengthened our marketing activities, analysing markets more precisely and improving our reporting procedures.

2. We set up our own production operations in the important markets, above all, increasingly taking the overseas distribution network under our own control.

These decisions were a major foundation stone for the logistic concept we use today. The establishment and expansion of an information system for our own distribution companies is today the basis for improved reporting on the market situation and competitors' activities. In our subsidiary companies, we have an improved control capability in terms of stock and sales situation, and the lines of supply both within national markets and to the world market are now shorter. Today, all these measures taken together constitute the preconditions for market-oriented production. They were one piece of the mosaic.

In assembly too, further pieces of the mosaic came into existence without being directly connected with the subsequent overall logistic concept. The reason was simply that we kept pace with developments in technology in order to modernise production and to increase the quality of our products, and as will be clear to everybody, these developments were going in the direction of control technology and the application of computers and microelectronics. At the same time, there was, of course, an effort to improve quality control.

We soon became conscious, however, that the pieces of the mosaic in the various areas of the firm were only half as valuable when seen on their own. They only become completely effective when fitted together and when they are integrated into an overall concept which aims at speeding up the flow of materials and assembly and which in so doing, brings the capability of the firm to deliver in line with the demands of the market.

To achieve this, it is necessary to have a procurement system with the necessary authority, and so in the case of our firm, the top management made the required organisational changes. An autonomous materials control system was set up, integrating the main purchasing and logistic functions. Working in close cooperation with the rest of the company, this system is responsible for the flow of materials all the way through from procurement to the despatch of products.

We see our main task as transforming market impulses into production in the most flexible and also the most integrated manner possible. But we do not see JIT only in production: Above all, we see JIT in the distribution channels and in the optimal use of warehousing. Product development is another area that receives impulses from the market that are as strong as the whole of the rest of production activities. That means that for us, watching the market and evaluating the competition are decisive impulses that are the source of our future orders, which new products must fill in the market-place as the market is also fed into development. From there, a mid-term development plan is established, which in the present case goes up to 1991 for the simple reason that it takes at least three years for a new product to go from the drawing board into production.

As far as the cooperation of the staff is concerned, what we did was to simultaneously inform everyone who would be affected by the changes (in other words, not just individuals) about what we proposed to do. We were quite open about how their tasks would change, and above all else we introduced appropriate training programmes so that those whose tasks had changed would be qualified to carry them out to their best ability in the new circumstances. You only have to do that kind of thing once because the word soon gets round in the firm that people can have a better, cleaner, and more interesting job if they are prepared to do the necessary training. And if you can do that right once, then there is not really a problem of acceptance any more.

You must take measures to ensure that your personnel are capable of working with the more flexible, and therefore more complicated, machines. In other words, you will in future have more skilled people in assembly than is usually the case today. That means moving away from employees who are just machine minders to specialists who know the whole machine and can therefore make full use of the quick changeover capability of the machine. That is just one aspect. That one cannot play fast and loose with manning levels in such a process goes without saying. Not only is labour legislation continually being extended, but so are other limitations, not only by the trade unions but also by the industrial tribunals. So flexibility as far as personnel are concerned cannot be taken for granted; it is something we must get used to. Therefore, we must build in flexibility with the machines: In other words, we must so arrange and equip our installations that we can produce ten different products with one installation and not two different products with five installations. Then we will be able to get the maximum benefit from the one installation, without it mattering so much that it is more expensive. But to do that, you will need more qualified staff and fewer unskilled staff: That is absolutely certain. It is a trend that is already visible.

In making these changes, we had absolutely no problems with the works council, and they in turn had no problems with the Metalworkers Union (IG Metall) because the changes, under the wage determination system that we have, enabled staff who were being upgraded in terms of extra skills to move up to a higher wage level. If more skills are demanded, staff must be paid more. So it is a development that is definitely in the interests of the workforce, its representative body, and the union.

Interplay of the Various Systems in an Integrated JIT Oranisation. Market Analysis and Sales Reporting: The Basis for Logistic Planning

In dealing with the role of the various functions in the organisation in formulating and implementing an integrated logistic concept, it should be made clear at the outset that JIT must include all parts of the organisation. The goal throughout must be to improve the ability to react and the flexibility to synchronise activities.

For a mass production company whose products must sell in competitive international markets, the logistic impulse must come from the market itself: Only what the market demands can or should be produced. Our company has therefore periodically extended its market analysis, and because we now have market analysis data that are more up to date and more accurate sales forecasts, we can carry out "rolling" marketing planning.

It is not only our subsidiary and associate companies that are linked into the reporting system but also the main general importers in the largest markets, the general agents overseas, and the second level of distribution in our biggest overseas market, the USA. Twice a year, at the time of the spring strategic planning and the autumn planning for the coming year, we have intensive talks with our foreign production and distribution companies about sales development. Consequently, from the sales side, we now have a much better basis for making the necessary dispositions in regard to procurement and production. We are now able to fit in more easily than before with the dispositions of our customers and can react more quickly to drastic changes in the market: We get fewer surprises. Sales data provide better conditions for supplying the market on an order basis and for implementing effective JIT management. All the same, we are not yet satisfied because communication can still be speeded up. So we are setting up a worldwide data transmission system that will allow us to call up any information necessary from our overseas establishments at the touch of a button in Waiblingen. We want to complete this decisive step very soon. But we are not considering transferring any of our design activities overseas. One reason is that we want to have all design activities in one place to be sure that we can guarantee the same quality all around the world in different production facilities and in order not to build just one specific model for only one market. Therefore, another very important task of our marketing group is to feed back information from the different markets that will allow our engineers to design a new product to make it fit all the different markets. Using this method, we can design and produce engines that can be marketed worldwide.

The role and importance of the end market for the establishment and functioning of a company's logistic system have been emphasised quite intentionally because for a mass production company, it is from this side that the impulse must come. The end market provides the basis for the successive dispositions in relation to the flow of materials and production. The information flow from the market to the firm is the part of the communication channel that gives the impetus. If market information only travels at a snail's pace or with the mail coach, reequipment of production facilities or punctual daily deliveries of components to be used in assembly alone achieve nothing: just as flexibility in production is useless when its benefits are wiped out by an antiquated and inefficient system for delivering the products to the customer.

Materials and Production Flow: The Main Artery of the Logistic System

The other part of the communication channel in the logistic system is the flow of materials and production, right the way through from procurement, via production, to distribution and the customer. For reasons of international competitive capability alone, the processing of orders with the aid of modern methods of data transmission and office communication has to be speeded up, and in this context it is no longer utopian to think of exchanging data and sending information on orders, deliveries, and payments from one continent to another without using paper.

The motor industry is already demonstrating what time reserves there are and what time savings can be achieved -- both in what it is doing now and in its plans for the future. Volkswagen, for instance, intends to reduce the average production time for their entire range of models from 33 to 15 days. A well-known example of JIT in the motor industry is that of Keiper-Recaro which, in time with Daimler-Benz's assembly line, delivers the seat fittings for a car by conveyor-belt every twenty minutes. Such a close connection means that the component maker is almost totally integrated into the production process of the end product.

Medium-sized firms, for reasons of competitiveness, also have to exhaust the time reserves inherent in a materials flow that is optimally controlled by a logistic system. This is because the longer the time spent on production in-house, the greater the danger of departing from market requirements. On the other hand, the shorter the routes from the producer to the customer, the smaller stocks can be that would otherwise be needed to ensure the firm's ability to deliver. At the same time, the need for warehouse space is reduced, which simultaneously results in lower investment requirements and savings on premises. The benefits are therefore numerous.

With a logistic concept, a high priority must be put on reducing the time of materials and production flows as well as on reducing all inventory, especially intermediate stocks and parts stocks. Materials represent a large part of costs and the

77

tendency is for their share of total costs to increase. A company's strategic goals must therefore be reduction in materials flow time and reduction of stocks so as to effect a considerable reduction in the materials link. Money tied up in floating capital in inventory should instead be transferred to investment capital, where it can be used for the rationalisation and automation of procurement, production, and delivery systems.

By adopting such measures, the overall warehouse turnover and the standard of service of product delivery can be appreciably improved. In our own company, the first successes came at the introductory stage of the new logistic concept so that, for example, the amount of the main company's capital tied up in inventory could be considerably reduced over the last three financial years.

This has encouraged us to continue along the way we have started, and we are currently introducing a programme-oriented logistic system based on a newly developed computer support system. This will enable us to improve the JIT system for the delivery of parts either bought in or produced in- house, as well as that for raw materials. In the near future, when we are able to control the inventory of the entire group from Waiblingen, we will have taken another decisive step forward. Also in the process of implementation is a so-called "market-oriented logistic system" which, within the framework of the world association of business electronic data processing users, will maximise the flow of information and materials from the manufacturer to the customer and vice versa.

Production Structure and Process: Decisive Factors in Company Logistics

Speeding up materials flow in the JIT sense and production based on orders are only possible when the necessary modern production structures are available. In plain language, that means production and assembly systems must be flexible and the plant must be increasingly (semi-)automated -- and that requires a large capital outlay.

At our own company, for example, several years ago we replaced old assembly lines for power saws with new flexible assembly systems that can handle different models. The new systems are also equipped with interposed automated work stations, and thanks to these flexible manufacturing systems, which cost ten times as much as the old lines, the set-up time for producing another model of saw has been reduced to 30 minutes. That is an immense step forwards in comparison with the previous situation, which was characterised by production limited to one model and therefore by long production runs, big stocks, and high capital requirements.

Today we need fewer people, and because we can produce in smaller runs, our production is more orderbased. We also need to carry fewer stocks to ensure delivery. In assembly, thanks to the capability to change over easily and quickly, we have attained the preconditions that enable us to react quickly to changes in market requirements in what is an important part of the production process. We are now on the point of introducing the assembly of sub-assemblies with fully automated changeover and the pre-assembled parts will be conveyed to the main production line in time with the latter. Our firm has now made great progress as far as JIT in assembly is concerned, but of course, one wheel in a gearbox by itself is useless -- all the wheels have to turn together.

A second example from our company shows even more clearly how the introduction of new technology makes the time factor in set-up times increasingly meaningless, while the possibility of more cost effective production in short runs becomes greater and the capability to react quickly to market demands is improved. This example is concerned with the full automation of the production and quality inspection of one part; a rail in three sections. To comprehend its significance, you must understand that the rail is also important in the spare part business, where even the smallest quantities are in demand.

By using computer-controlled warehousing and automated changing of the heavy stamping tool, the set-up time, which a few years ago took two hours, has been reduced to a mere three minutes. This completely automated changeover is integrated with a conveyor-belt installation with automatic changeover for raw materials (strip steel -- steel coil) and automatic feeding into the machine of the steel plates out of which the rails are stamped. Combining these two fully automated processes makes the production of the rail completely automatic.

Thanks to automation, which at the same time implies a consistency of quality, it is now possible to begin production of a different type of rail five minutes after the last one. So, for instance, when we receive an electronically transmitted order from the USA for a few spare rails, we have the technology to make the rails one hour later and to send them the same day by air freight to the USA. The customer can get the spare part he needs on the following morning. Previously it was impossible to deal with small orders for spare parts because of the high cost of set-up time or changeover. But, here again, it is only through integration into a comprehensive logistic concept that the advantages of advanced technology can be realised.

A high disturbance factor in production always used to be the unacceptable quality of material and of finished or bought-in parts. With JIT, this factor grows considerably in importance, and for that reason also, the importance of quality inspection and assurance becomes greater.

In the third example from our firm, a better approach to quality, automatic quality inspection, and other improvements in the production system that this led to produced measurable results: The costs for scrap and reworking in the production of one very difficult part fell from 72 percent to just 16 percent of production wages. In money terms, this was a saving of DM 1 million. The saving in time was 42 percent, and the saving in inspection costs was 32 percent. These figures speak for themselves and demonstrate how energetically we have tried to deal with the disturbance factor of "scrap" in the area of production. In so doing, we also improved the conditions for implementing a more smoothly functioning JIT system.

Establishing a JIT system -- by which I mean an integrated total system, starting with assembly and going right through to the market -- one has to think of a time span of ten years, and I do not mean that instant success can be attained in only one area. The delivery of one part in time with daily production scheduling, for example, is not very significant if the finished products are sitting around in the loading bay for a fortnight waiting for transport. The whole idea only makes sense when it is integrated into a comprehensive system, for which a number of preconditions are necessary. Creating these conditions is relatively difficult in many firms, and it does cost money of course. So they have to be implemented on a longer-term basis, but companies that have so far done relatively little to create the first pieces of the mosaic would be well advised to speed up the process a little so that they can get the whole picture sooner.

Role of the Supplier

Naturally it is not enough to establish the basis for a JIT system only in one's own firm, even if this is the first and essential condition. The whole thing will remain patchy unless the suppliers, especially the main suppliers, are brought into the concept. If the system is to function properly, greater demands must be put on the suppliers to make deliveries "in time" and to assure the quality of components etc.

Like the manufacturer of the end product, the main suppliers must have an adequate logistic capability, and they should also be connected to his data transmission system. Shorter material flow times demand synchronised and precise delivery times, and delays in delivery must not occur because of the immense disturbance they can create for the flow of production.

What applies to delivery times also applies to the same degree to the quantities ordered, as a shortfall soon creates the same problems as failure to meet delivery schedules. Unacceptable quality likewise has the same negative effect since poor quality is an extremely serious disturbance factor in production.

Such demands confront main suppliers with some problems, but the motor industry is now showing us the way; and anyone who does not meet the demands and does not adapt will not get any more orders. Even medium-sized batch production companies like our own must make such demands on our suppliers to be competitive internationally. We believe we have solved the problem for a medium-sized company relatively well.

We began to hold regular meetings about quality control with our main suppliers several years ago with a view to getting rid of the duplication in inspection that was done at the suppliers' before despatch and at our company at goods inwards. That can very easily be achieved by making cards with the rules for quality control, keeping records over a longer period and seeing what the quality delivered is. When, over six months or a year for instance, reject rates are less than 1 percent or 1/2 percent, then you can agree to accept direct deliveries without inspecting the goods again at your own site.

The second step was to institute a so-called suppliers' day, to which we invited all our most important suppliers. We explained our philosophy to them and what we were doing and what we were aiming at. We suggested to them that they should participate in this system and informed them that although we were making considerable demands, we were also prepared to make two-year or three-year contracts with them so that the investments they would need to make would not be wasted. Otherwise you will find that unless your company has the demand power of a major automobile manufacturer, for instance, suppliers are not particularly inclined to make the investments you require.

This strategy was successful, and at present we insert about 25 parts directly into the assembly process. They are delivered in time every day or every other day, and it is our intention to increase the number of such parts to something like 120.

One important precondition for the functioning of such a direct delivery system is the intelligent application of electronic data transmission systems that allow you to say with much greater precision what part you need and when you want it. If JIT only meant pushing our stockholding problem on to the suppliers, then that would result in higher prices for us. In practice, we have found that some buffer stock must be held somewhere, and as a rule we are the ones who hold it. This is also on the grounds of flexibility as far as changeover in assembly is concerned, so that the demand is mainly on the supplier's internal logistic system, rather than on his carrying larger stocks to be able to ensure reliable delivery.

Conclusion

JIT can only be introduced by convincing people and not just by issuing a directive from above. We ourselves first took the road of applying solutions to particular areas, starting with the individual pieces of the mosaic of which I spoke. Then, after lengthy discussions rather than in a flash, we came to the point, of deciding that if we really wanted to get the full benefit of JIT, we would really have to link all the different activities together to make it work all the way through from the supplier to the customer. This process took place in our top management meetings and lasted quite a while. But if you are in a position to listen objectively to counter-arguments, to discuss them, and to bury them together over a period of time, then you can in the final analysis come to a complete consensus at the top management level. That is one of the first and most important hurdles to get over as each area must stand behind this overall conception with complete conviction. When each area runs a pilot project and assumes responsibility for informing colleagues and dealing with their reservations, for training them, and finally for convincing them, then you can succeed in gaining full acceptance of the system down to the last operator.

At the present stage of technological development, a company can only be competitive if it is able to react quickly and flexibly to the demands of the market and at the same time use all the means at its disposal to rationalise material flows. In addition, it must, among other things, use microelectronic technology to bring all the functional areas of enterprise into a single integrated logistic system organised on the basis of the JIT approach. In so doing, it is possible:

1. to coordinate the flow of information and communication betweeen procurement, production, and distribution in both directions and

2. to speed up the material flow, right through from procurement, via production to the customer, by using modern data transmission and office communication systems, and to ensure that it fits in with production schedules and

3. in such an integrated system, quality occupies the position of top priority, and price reduction at any cost is not the priority. Quality must be in first place, and then it can, within bounds, command the price that it costs. It cannot be done any other way.

JIT IN LARGE COMPLEX COMPANIES

Siegfried Höhn
Volkswagen AG

When we speak of JIT, we are talking about a management philosophy that has been superimposed like a kind of ideology on a management system. The complexity of the JIT problem arises not from the philosophy itself but from the application of the means we need today to put it into practice. We talk about the complexity of flexible machines, the complexity of demand, the complexity of manufacturing processes, and we talk about automation and microelectronics: all of which make up a multiplicity of elements of supply and demand that are at the same time connected with this philosophy. As far as corporate strategy and competitiveness are concerned, the first question is whether such a philosophy holds out any advantages for the company in the context of its competitive situation. And if we look at JIT as the opposite of the company's functional organisation based on the priniciples of Taylorism, then it is a reversal. The word "bonding" rather reminds one of the chain of a machine, but in this case it is not just mechanical: What is meant is the cooperative linking together of departments, operations etc. Here each company must determine for itself what its state of development is in regard to competition. Inevitably, within the company, there will be a certain amount of repetition of experiences that have already occurred elsewhere, which may justify the use of the word imitation, but each company must find its own solution. If you look at large complex organisations, of which automobile manufacturers are an example, then you can see that the JIT philosophy itself is a powerful challenge. In the case of Herr Stihl's firm, which by German standards would be classified as medium-sized, it is certainly easier to handle JIT. This is because all of management can be reorientated towards such an autocratic JIT philosophy by the proprietor (and managing director). In large complex organisations, greater readiness on the part of all involved is required, and the problem with the Tayloristic splitting of the task into separate functions is the heart of the matter. In a large firm, ensuring survival in one's own function takes precedence over the success of the enterprise as a whole.

Now, bound up with this question are a lot af administrative considerations, such as how to motivate managers to try JIT when to do so means that their own function might be downgraded; when the scope of their own job might be reduced; and when success might be apparent not in their own function but somewhere else in the enterprise. How, then, in the latter case of a person who is integrated into this bonding of the organisation's various units, should his participation in the profits be arranged? These are questions that affect motivation, and therefore, the larger the organisation, the greater is the complexity of the challenge inherent in implementing JIT; though as far as Volkswagen is concerned, that does not mean the challenge is not being met, in spite of the complexity of the motor industry. This is not because the Japanese did it before we did but simply because we believe it is necessary in order to use capital more efficiently and to improve productivity in the context of legal restrictions on working hours. Moreover, to be able to meet demand on the basis of orders so we can deliver any model the customer wants within a specified time, there is no choice for us but to set up linked units with heavy information loads, which over a period of years bring about throughput more and more in the sense of bonding. That is a very long, drawn-out process, which in complex systems can only be completed with heavy information support. It slowly achieves its purpose by bonding together this pool of information and the physical activities that go with it; including the change in the organisation that is very slowly introduced at the same time. If the latter is attempted too quickly, there will be a lack of sufficient understanding from the point of view that in large firms, the role of the unions and, in the German case, of the works councils, makes it imperative to discuss each step very carefully, to vote on it, and to lay down precisely what the effect on manning levels will be. Imitation, in the context of JIT, is a very complex question, but it seems to me that our job is to find solutions that are tailormade for each individual enterprise.

One problem I am often asked about is the significance of accidents for JIT. Of course, there have unfortunately been accidents and interruptions in the supply chain. That may seem to be a contradiction of JIT, but any such rationalised system must take actual operational conditions into account. That is the risk inherent in the central planning of throughput and delivery times, which must be able to deal with unforeseen occurrences. The solution can take different forms, and in this respect, there certainly is a great difference between European and Japanese companies. Take, for instance, the principle of having several suppliers who can be requested to cushion this or that shock.

For companies competing internationally, there is always the opportunity of a tighter alignment in favour of greater productivity on the part of the supplier, and the risk of interruption can often be offset by the capability that allows it to have the same part produced internationally, either in direct or third party collaboration, and flown in by air freight. In any case, one has to say that interruptions are a part of life. Sometimes one may greatly regret the reasons for them, but there must be preparations for such contingencies in any logistic system. This problem can be dealt with.

A TRADE UNION VIEW ON JIT IN GERMANY

Walter Riester
IG-Metall

JIT means greater involvement of personnel in all areas, including problem-solving. JIT has aspects that concern raising productivity, but the primary goal - let me put it crudely - is not to reduce the number of heads but to use heads to innovate and to improve the overall system. JIT is not a value analysis of fixed costs but another way of distributing tasks. JIT only works properly when job enlargement, job enrichment, and other parts of the humanisation of a work programme have been achieved.

Even if one agrees with this analysis, I am afraid that it does not, as it stands, correspond with what happens in practice at the company level. And let me say plainly that in Germany, I hear more talk about JIT than I see being practised at company level. This has the danger that JIT is imitated and that these basic principles, about which there has been so much discussion, are not seen to be integrated into the overall system. Looking at the basic principle, our point of view is that JIT presupposes a higher standard of qualifications among employees. JIT presupposes a high degree of cooperation on the part of employees. That being so, I wonder why these questions about increasing demands for qualifications and for employee cooperation are so little discussed within companies, while logistic processes are discussed so much? I entirely agree that it is not enough to implement JIT at Volkswagen without bringing in the suppliers; but, however correct that is, it also means I cannot introduce the JIT principle at enterprise level without reflecting that in the final analysis, it is the people engaged in production who implement it. This is not only from the viewpoint of qualifications, by which I also mean, for example, understanding of the system. I have some acquaintance with the circumstances of Herr Stihl's factory, which are not typical, and so I can imagine that in such a case, it is possible to implement the JIT principle more quickly. But I would warn people against thinking it can be done right across the board, without first having done the preparatory work with the employees. As with the preconditions, I can agree entirely with Herr Stihl's point that things must be settled at the level of the individual segments. And when they are settled there, then I would agree that gains in productivity do not simply originate in the system but out of the optimal use of each of its elements. These include, for instance, the optimal qualification of the staff and the optimal coordination of EDP and the flow of production.

Those are elements of productivity, and as a union we see both opportunity and risk in them, depending on how they are deployed. Productivity offers us the opportunity of producing larger quantities of goods and services in the near future and of reducing the working hours of employees. But if one is going towards a system that has a high degree of employee acceptance as an essential precondition for it to function at all, then I think such a task can only be tackled by everyone together and can only be approached as an opportunity. On the other hand, if one takes risks and tries to force it through, then I do not think it will work at all as far as employees are concerned.

Another important aspect is flexibility in regard to time. I rather regret that the question of flexibility in the sense of time has been so preponderant in the last three to four years. It is important, but it may detract from the concept of flexibility as a whole.

We have indeed a lot of flexibility in wage agreements in Germany. We have the regulation of overtime, short time, shift time, preparatory work, refinishing operations, and part-time work. In other words, we have a whole range of possibilities for departing from the normal full-time employment relationship in prescribed individual cases.

We know that the metalworking industry in our area for example would like regulations to go still further, preferably in the form of an agreement in which there are no limits at all, so that the particular conditions relating to workplace agreements can be laid down outside of regular working hours. Something like this was put forward in the wage negotiations in 1984 and in further negotiations, but naturally we do not want it. What we want in an agreement arising out of a wage agreement is a definite framework for working hours, which can then be filled out in a similar way to the one we use now.

But I would like to point out again that we are detracting a bit from the view that flexibility has a lot to do with qualifications. Recently we brought in a proposal from Herr Stihl and the metal industry's employers' association. This was within the framework of wage negotiations, and we said, also from the standpoint of our own interests, that we would take up the two concepts of qualifications and flexibility. Let us make a wage agreement, for instance, so that the parties at plant level can bring together, according to the needs of the plant, particular areas where tasks that require different qualifications but are connected are carried out. Let us make an agreement covering wages, so that employees in a particular area for a particular period of time have a claim to further qualifications, which will then enable them to carry out all the tasks in that area. That could be the area of the senior foreman (Meister), for instance, or it could be an area running right through the plant. In other words, it does not have to follow plant organisation rigidly. Then we would have the preconditions in terms of qualifications, and tasks of different kinds could be carried out by the employees.

So we see a major addition to flexibility in higher qualifications. And I would like to give a slight warning, although working hours are certainly one aspect of flexibility, against discussing and pushing for flexibility exclusively or overwhelmingly in terms of time, as has been the case in the last two or three years. It rather detracts from the view that flexibility has a lot to do with better qualifications.

A COMPARISON FROM A JAPANESE POINT OF VIEW

Takahiko Amanuma
A.T. Kearney GmbH

If I try to make a comparison in terms of efficiency and per-haps also management concepts, including the concept of invol-ving employees in problem-solving, I want to stress that mid-dle management is very important. I even think this is the key point. There is a great difference between Japanese compa-nies, especially those that have had success with JIT, and German companies. In German companies there is a section chief responsible for each production section, and his responsi-bility is different from that of a section chief in Japan. Their evaluation is different.

In Germany, or perhaps Europe, each section tries to achie-ve the maximum result for the section by making 100 percent use of the expensive machinery and by producing more pro-ducts. If the line is stopped, then the superior will upbraid the section chief. But in Japan, the company as a whole is regarded as important. The optimum performance of the com-pany or the plant, or the degree of a particular section's con-tribution is the criterion for evaluating the section chief. This is not only a question of difference in motivation: It is per-haps a difference in overall policy in German companies. In fact, each section is not so independent in seeking optimum performance in increasing productivity. If you produce more products for the section, the contribution may appear to be greater, but perhaps that amount is not needed at the next stage of production. It may indeed often hinder the work of other parts of the system and of the plant as a whole.

Discussing JIT in an international context, it has been alle-ged that large Japanese companies like Toyota and others were exploiting their sub-contractors and that this was the reason why they were so successful in world markets. The answer given by Toyota and companies that had introduced JIT sys-tems was that they did have firm links with sub-contractors but that they did not just shift their inventories on to them. They stated that they also passed their know-how and control know-how to their suppliers so that the latter were then enabled to use JIT systems. In spite of a certain lack of skilled labour, they are in fact given help by the larger companies so suppliers can in turn reduce their own inventory. And they can design and control their systems better than in the past. Of course, these reduced inventories are then shifted on to the next stage down the line, and the next stage, and the next; and at each stage everyone has to try to improve his or her own methods.

There remains the specific question as to whether there might not be some inherent limitation on competition in having such a very close relationship between the manufacturer and the sub-contractor. It does look like it on the face of it, I must admit. But if the term "free competition" or "free enterprise" is not just limited to supply and demand, and if it is interpreted a little more liberally, namely to include the offer of control know-how and better know-how on the part of the manufacturer, then suppliers benefit. They will look carefully at the companies to see which of them offers the best know-how. It is not just the price that is involved but also the possibility of growing together. That is another aspect that has to be taken into consideration.

CONCLUSIONS

Horst Wildemann
Universität Passau

JIT is certainly an important element in production strategy, and its goals (such as production according to market requirements, increased flexibility through short production times and short changeovers, realisation of the potential for rationalisation, and perhaps the concentration of all the activities of the enterprise on the actual process of adding value to the product) have not always been striven for in a very efficient manner. But I do not mean that JIT is the only production concept or that there is no alternative.

There are alternatives, for instance, along the lines of computer-integrated manufacturing. JIT is made up of many building blocks and it seems to me that one of the most important is integrated information processing throughout the logistic chain. And we should not allow one of the great strengths of European firms (compared to Japanese and other competitors), namely the penetration of electronic data processing into the decision-making sphere, to be put back by a concept that originally developed without EDP support. That cannot be the point. It is something that we must discuss further.

By contrast, another idea seems to me to be important, and I would like to take it back to a fundamental scrutiny of the activities in the logistic chain. Since F.W. Taylor, we have assumed that the way to get maximum efficiency lay in strict separation of decision-making from implementation or application and in handing each activity over to specialists. But here, an alternative concept shows us that, perhaps because of Tayloristic theory, the critical mass of specialisation has been somewhat exaggerated and that one can actually increase efficiency by reducing specialisation.

Added to this is another important idea that also comes from the Tayloristic approach, namely the central coordination and control of all activities in the logistic chain. In the JIT concept, this is removed in favour of decentralised decision-making with the conscious involvement of the ordinary person.

Let me draw a macro-economic parallel rather like supply and demand, illustrating how to improve coordination within the enterprise.

It was plain from several of the presentations that the JIT concept has a decisive influence on the organisational structure, which becomes flatter, and in which people are concentrated into smaller organisational segments. In each of these segments, there should be an entrepreneur, rather than an administrator who coordinates the activities. A member of the board at Siemens, for instance, put it this way: You have to become small in order to grow -- and this is only an apparent contradiction.

Relations with suppliers have already been thoroughly dealt with, and it is incontrovertible that in these relations the power of demand plays an absolutely decisive role. But I can also imagine that the starting point in building up a relationship with a supplier does not preclude continually asking oneself whether the partner is equal or basically the weaker link in the whole chain: However, I also think that cooperation is definitely possible within the framework of a total increase in efficiency. Here I am thinking solely of a long-term contract basis within which there are higher volume deliveries. I am also thinking of the incorporation of the development and product development capabilities of the supplier into the end product. I am thinking of the ordering systems, which are certainly paperless, as we have already discussed. And I am thinking of one of the most crucial goals of JIT, which is the firm's concentration on the actual process of adding value to the product. That also means firms taking over functions as specialists: for example, taking over transport and warehousing from haulage firms, as has already happened in several cases. I am, of course, aware of the objection that stocks are really just being pushed down to a lower link in the chain. From a business point of view that already has a considerable effect, as it means products are held in stock at a lower level of value added; but from a JIT point of view, that is not the right concept -- it must be a temporary phenomenon. In my view, and this is something I have observed in the U.S., the JIT concept has another effect on suppliers. Suppliers in geographical proximity and with smaller production units are preferred. That implies bringing back employment to a given economy.

Let me mention one aspect that corresponds with my observations here in Europe. It is the declared aim of many companies to reduce the number of suppliers per part, although single sourcing is not their declared intention. The rule is for a distinction to be made between main suppliers and subsidiary suppliers.

From this, it is evident that JIT has an effect on competition in the sense that it changes competition. Competition refers to entry barriers to the market, and many large companies that place orders naturally try to bring this competition into the process. Here, elements that have already been tried in Japan are used. There is, for instance, the possibility of making costs transparent by continuous value analysis and quality audits and by getting proposals on how the implementation of the organisation's plans should be carried out. There is also the possibility of, for instance, applying the insights of group experience theory to organisation- building. A smaller number of suppliers when quantities remain the same means larger volumes for them, and that in turn has a continuous rationalising effect in which the company placing the orders must also participate.

In the final analysis, such dual sourcing means that if the supplier's existence is threatened, then the customer company must take it over financially.

It was also said earlier on that firms are establishing production units near the market, though basically only for a part of their productive capacity and, only in the rarest cases, development capability as well. But here again, it seems to me that the general evidence is that JIT delivery is not suitable for each and every part or for each and every area: It has its own special area of application, and there it is efficient.

Now let us deal with the third building block, which is one that seems to me to be extraordinarily important, namely the reorganisation of production in one's own company. Here the highest goal is to decentralise capacity; in other words, to alter the layout. The highest goal is to change throughput times from the standpoint of maximisation. The motto cannot be to minimise floating capital. The motto can only be to maximise the efficiency of both investment capital in investment in machinery and floating capital. This is because a one-sided reduction in floating capital would in time lead to a disproportionate investment of capital in capacity, which, as is well known, leads to a considerable burden of fixed costs.

This has effects on a company's investment policy and also implications for the optimal size of a company unit. The optimal size of a company is now not determined by unit cost comparisons but by evaluation of the whole logistic process and the harmonisation of all capacities in one integrated logistic chain.

One other important point is JIT and the management concept that stands behind it. I think many people look at JIT in the same way as the seven blind men who each felt one part of an elephant and then tried to describe the whole of it. Of course JIT has many aspects, but the only way to give it a strategic direction is to see it as a whole. It should be made clear that the concept cannot be implemented from one toehold alone and that it is better to approach it through a number of pilot projects that can then be integrated with each other. In other words, it is more a concept of learning by doing. Smaller organisational units, decentralisation of decision-making, and management at group level all imply a different style of communication and different management principles. In the U.S., they talk about hands-on management, but here in Germany we do not at present have such a vivid expression.

JIT implies changes in management style, which is certainly different in Japanese companies from European or U.S. companies. Now, one might say that JIT principles can best be implemented in German firms when both expert knowledge and competence, or competence both in knowing and implementation, are found in the management. The JIT concept, to put it briefly, is best carried out by convincing people rather than by giving them orders. What I mean is that up to now, managers have been orientated towards the maximisation of a particular function, such as purchasing or production -- and great strides have been made in maximising performance on a functional basis. But the important thing with JIT is the overall maximisation of performance of the whole logistic chain, and to me it seems that convincing managers of this is indeed a difficult task.

Top management often accepts the JIT concept very quickly because published case studies show how advantageous the system is and because it makes competitive sense: As Herr Stihl said several times, it is a necessity. But if we look at the level of implementation then, as we heard from the Amalgamated Engineering Union, the situation is different. In Germany we have had a project to humanise work, and its primary objectives have been job enrichment, job rotation, and further training: all of which are necessary in JIT. More highly qualified people are needed, and so these programmes go in parallel with the potential for rationalisation that they bear. JIT also has implications for reward. I see greater problems of acceptance on the part of middle management, and Mullard Ltd. has shown how to deal with these problems and to bring things into line strategically. These problems have readily understandable roots because for each individual manager, JIT means an increase in personal risk in the fulfilment of his task. Take the case of a purchasing manager, for example, who always stands between the two poles of reliability of supply and cost effectiveness. He can get reliability of supply by ordering the same part from several suppliers; and cost effectiveness by ordering larger quantities from each supplier -- but remember, stocks are necessary.

Or take the case of a production manager who would like to have continuous stable production not coupled to sales, so that he can sort out the problems. For him, small production runs and permanent change cause these problems and also costs. Think of the sales manager who defines a shorter and shorter delivery time than that of production item plus resupply time. Remember that stocks are necessary for such a strategy. This means that a thoroughgoing change in behaviour is necessary for the whole middle management in order to introduce JIT. I see a major difficulty in bringing about changes in understanding problems, problem-solving, and communication.

We have carried out investigations to see whether such a chain as there is with JIT, when equipped with self-regulating control mechanisms, regulates supply and demand from the users' standpoint at any given time. These mechanisms were simulated in an opposite way from centrally coordinated control mechanisms that control a whole logistical chain by instructions and feedback. The fact is, the simulated results show that the principle of decentralised self-regulating control mechanisms is more subject to interruption than a centrally directed system in the logistic sense. In other words, decentralised systems are exposed to greater risk, but risk-taking itself arises out of supply and demand and out of competitiveness. When a fellow competitor takes these risks and also employs preventive strategies of risk avoidance, then he has a competitive advantage if he can do that with a lower amount of floating capital.

Let me just make a few more points about the business implications. There is no question that by applying the JIT concept, companies can reduce inventory, increase flexibility, and sometimes to their surprise, increase productivity by as much as 20 percent or more. From a theoretical viewpoint, it is correct to say that only in a limited way are these increases in productivity due to the JIT concept and that they proceed more from new ways of thinking things through and of coordinating all resources in a strategic manner. As a result of this competitive advantage naturally follows. The question then arises as to whether, from a business point of view, it makes sense to implement such an approach. As far as I am concerned, the only answer to this question is yes.

My main thesis is that a simple transfer of this type of production system into a European setting can at best enable us to keep up with the conditions of competitiveness that some other competitors have already attained. As I see it, the intellectual task here is to apply techniques that will enable managers of a company to attain perhaps a small competitive advantage over their rivals, and this cannot be done by imitation but by innovation -- including innovation in the organisational field.

JIT is undoubtedly a logistical and organisational innovation from Japan that we are now meeting. It is an alternative to previous ways of doing things in Europe and an alternative to other concepts connected with, for instance, computer-integrated manufacturing. If one were to give a brief view, one might say that the next step towards increasing efficiency within a short space of time would be the implementation of JIT, not in its pure form, but in a form adapted to the needs of individual companies. This would give them competitive advantages and would enable them to earn the money needed for the next steps. In my view, the second step would be the realisation of computer-aided manufacturing, and the third would be flexible manufacturing. The aim would, of course, be to satisfy the needs of the customer better; perhaps also with the ulterior motive of attaining greater competitiveness.

APPENDIX I

APPLICATION OF JUST-IN-TIME
MANUFACTURING TECHNIQUES IN THE UNITED KINGDOM

Professor C.A. Voss and S.J. Robinson

SCHOOL OF INDUSTRIAL AND BUSINESS STUDIES
UNIVERSITY OF WARWICK
COVENTRY   CV4 7AL

The Concept

JIT may be viewed as a production methodology which aims to improve overall productivity through the elimination of waste and which leads to improved quality. In the manufacturing/ assembly process JIT provides for the cost-effective production and delivery of only the necessary quality parts, in the right quantity, at the right time and place, while using a minimum of facilities, equipment, materials, and human resources. JIT is dependent on the balance between the stability of the users' scheduled requirements and the suppliers' manufacturing flexibility. It is accomplished through the application of specific techniques which require total employee involvement and teamwork.

The Elements of JIT

The fundamental aim of JIT is to ensure that production is as close as possible to a continuous process from receipt of raw materials/components through to shipment of finished goods. Some of the elements of this system are as follows:

- Flow/layout
  The physical layout of the production facilities is arranged so that the process flow is as streamlined as possible; i.e. for each component, the proportion of value-added time is maximised, the proportions of queueing and non-value-added time are minimised. The flow is analysed in these terms and the layout configured accordingly, resulting in the reduction and/or elimination of stores and conveyors; U-shaped or parallel lines may be applicable.

- Smoothed line build rate
  This should be consistent with the process flow, i.e. as smooth as possible. JIT systems often try to smooth the build rate over a monthly cycle.

- Mixed modelling
  A JIT objective is to match the production rate to order demand as closely as possible. One method of doing this is to increase the flexibility of production lines to allow the concurrent assembly of different models in the same line.

- Set-up time reduction
  The object of reducing set-up times is to enable batch sizes to be reduced. Some companies aim to achieve a batch size of one.

- Work-in-progress (WIP) reduction
  In JIT, WIP reduction is used to highlight production problems previously shielded by high inventory levels; these have to be resolved without delay in order to maintain production.

- Kanban
  Kanban is a pull system of managing material movement comprising a mechanism which triggers the movement of material from one operation through to the next.

- Quality
  The achievement of high quality levels is a prerequisite of successful JIT. Commonly used quality programmes in support of JIT include zero defects, statistical process control and work team quality control.

- Product simplification
  This can be achieved by two measures, the rationalisation of the product range and the simplification of the method of manufacture by, for example, using fewer and common parts.

- Standardised containers
  The use of small standardised containers to transport components.

- Preventive maintenance
  Effective JIT requires the removal of causes of uncertainty and waste. A major cause of uncertainty are breakdowns; JIT programmes are normally supported by preventive maintenance.

- Flexible workforce
  In order to support the objective of matching the production rate as closely as possible to order demand, flexibility of all resources is required; this applies in particular to labour. Cross-training of the workforce is one mechanism often used.

- Organisation in modules or cells
  Many JIT factories are organised in small autonomous modules or cells, each cell being totally responsible for its own production and supply of the adjacent module. Within the cell the workforce is trained to work as a group, and often many activities normally considered as staff functions such as scheduling and maintenance are brought into the cell or module.

- Continuous improvement
  JIT implementation is not a one-off effort; it embodies the ethic of continuous improvement, which needs to be supported by all levels of staff in the production team.

- JIT purchasing
  Materials and components are purchased in compliance with well defined requirements in terms of quantity, quality and delivery. The success and resulting performance reliability of this system is based upon co-operation between the purchaser and supplier.

APPENDIX II

JUST IN TIME SYMPOSIUM
COLOGNE, 10-11 SEPTEMBER 1986

Chairmen, Speakers and Panel

AMANUMA, Takahiko
  A.T. Kearney GmbH, Management Consultants, Düsseldorf

BAYLISS, Peter
  Materials Director
  Toshiba Consumer Products (UK) Ltd., Plymouth

BREIDBACH, Hans-Josef
  Stellvertretender Direktor
  Institut der deutschen Wirtschaft, Köln

CURE, Kenneth
  Executive Council Member
  Amalgamated Engineering Union, London

EDWARDS, Duncan A.
  Marketing Manager
  Consumer Division, Mullard Ltd., London

HÖHN, Siegfried
  Bereichsleiter
  Volkswagen AG, Wolfsburg

HOLL, Uwe
  Mitglied der Geschäftsführung
  Institut der deutschen Wirtschaft, Köln

ISAKI, Shoji
Deputy Director General
Japan External Trade Organization (Jetro), Düsseldorf

RIESTER, Walter
Bezirkssekretär, Bezirksleitung der IG-Metall
Stuttgart

STIHL, Hans Peter
Geschäftsführender Gesellschafter
i.Fa. Andreas Stihl, Waiblingen

TREVOR, Dr. Malcolm
Director, Japan Industrial Studies Programme
Policy Studies Institute, London

VOSS, Prof. C.A.
School of Industrial and Business Studies
University of Warwick, Coventry

WIENER, Dr. Hans
Project Director
Anglo-German Foundation for the Study of Industrial
Society, London

WILDEMANN, Prof. Dr. Horst
Lehrstuhl für Betriebswirtschaftslehre mit Schwerpunkt
Fertigungswirtschaft, Universität Passau

USAGOU

DISCARD

Malcolm Trevor, Editor

The Internationalization of Japanese Business

European and Japanese Perspectives. Proceedings of
the Second Conference of the Euro-Japanese Management Studies Association
1987. 2o9 Pages. ISBN 3 595 33816 5 (Campus Verlag)
ISBN 0 8133 0531 4 (Westview Press)

Campus Verlag / Westview Press